VOLUME 35

MERLIN-POWERED SPITFIRES

BY KEV DARLING

specialty press
PUBLISHERS AND WHOLESALERS

Published by
Specialty Press Publishers and Wholesalers
39966 Grand Avenue
North Branch, MN 55056
United States of America
(800) 895-4585 or (651) 277-1400
http://www.specialtypress.com

Distributed in the UK and Europe by
Midland Publishing
4 Watling Drive
Hinckley LE10 3EY, England
Tel: 01455 233 747 Fax: 01455 233 737
http://www.midlandcountiessuperstore.com

ISBN 1-58007-057-4

Printed in China

Front Cover: *Built as a Spitfire LF.IX this aircraft, MH434, was credited with shooting down at least one victory, a FW-190, whilst in use by the RAF. Currently the Spitfire sports the temperate colour scheme and the codes of No. 222 Sqdn.* (Danny Jacquemin)

Back Cover (Left Top): *Wearing the camouflage scheme associated with service in the Middle East and the XT codes of No. 603 Sqdn, Spitfire Mk.II, P7350, banks towards the camera. Noticeable on the wing upper surface are the reinforcing strips above the undercarriage bays.* (Dave Stewart)

Back Cover (Right Top): *For use in the Middle East the Spitfire required extra filtering to keep the amount of sand particles to the minimum and thus reduce damage to the engine.* (Chris Michell)

Back Cover (Lower): *This beautiful shot from the rear of Spitfire AR501 reveals the strengthened wing root walkway and strengtheners on the wing upper surfaces.* (Dave Stewart)

Title Page: *Awaiting delivery from Castle Bromwich to its first operational unit is this Spitfire LF.IX which sports a pair of cannon and four machine guns for that particular role.* (C P Russell Smith Collection)

Table of Contents

Merlin-Powered Supermarine Spitfires

INTRODUCTION

AND ACKNOWLEDGEMENTS

The threnody of sound builds to a minor cresendo as the speck seen in the distance grows in size. As the small shape begins to grow the slender shape becomes more defined. The growling whine reaches a peak as the aircraft we have been watching becomes more defined. A quick flick of the ailerons and the elliptical wing shape is revealed before the sound fades on the wind and the fighter is gone.

Go to any airshow and amongst the warbird fraternity there will be airframes powered by the Rolls-Royce Merlin engine. Some will be Mustangs, possibly a Lancaster, even a Hawker Hurricane or two, however it is highly likely that the carrier of the engine will be a Supermarine Spitfire. Each type has its own particular soundwave, and that of the Spitfire is not only distinctive but a source of great pleasure to its many fans. It is also the fighter that characterised the British effort in the Second World War (although fans of the Hawker Hurricane will no doubt disagree.) The Supermarine Spitfire is today seen as one of the greatest fighter aircraft ever.

It owes its conception to one man, R J Mitchell, and to a series of events including the Schneider Trophy races. Although the aircraft that Supermarine would develop in later years bore no direct relationship to the S.6 seaplane racers, the concept behind the streamlining of the airframe and the shoehorning of the largest powerplant possible into the smallest space proved that such an approach was feasible. As these events were happening, the Royal Air Force was still equipping its front-line squadrons with such aircraft as the Bristol Bulldog, the Hawker Fury, and the Gloster Gauntlet and Gladiator series of biplanes.

In Europe one man was in the process of rising to power in Germany and his influence would force that continent into a war that would eventually become almost global in its scope. Against this backdrop the Spitfire would be born. Small, fast, and agile it would be churned out in vast numbers to equip the squadrons of the RAF and its allies. In the seminal Battle of Britain it would share combat honours with the Hawker Hurricane which would in fact claim more victories than its Supermarine rival.

As the war would progress, so the design would develop. Clipped, clapped, and cropped would be the phrase used to describe the Mk.V converted for low-altitude work.

Developments would also see Spitfires operating at the other end of the spectrum as pressurisation was incorporated so that the fighter, complete with modified engine, could chase the Luftwaffe's high-altitude raiders.

Further changes would see bombs, fuel tanks, and even kegs of beer loaded on to the airframe whilst some daring souls would streak across Europe on reconnaissance missions in Spitfires painted, of all colours, pink.

The aircraft also went to sea where it equipped the squadrons of the Fleet Air Arm aboard escort and main fleet carriers. This, then, is the story from a slightly different angle of a fighter that still graces museums and airshows, where the growl of that famous Merlin engine can still be heard.

As always a work such as this is a multi person affair so I would like to thank the following for their much prized assistance: Damien Burke, FAA Archive; Nick Challoner; Dave Stewart; Danny Jacquemin; Chris Michell of Airframe Assemblies; Ray Deacon; Owen Morris; and Sander Wittenaar. Special thanks must go to Peter Russell Smith who yet again came up trumps with those essential photographs; and that doyen of Spitfire historians, Eric B Morgan, who also came to my rescue with those rarities once again.

At the end, but always in the centre, are the team at Specialty Press and my good friend and designer, Dennis R Jenkins, without whose sterling efforts this slim volume would not exist.

Kev Darling
Wales
September 2001

STARTING 1 THE RACE

THE EVOLUTION BEGINS

One of the greatest disasters that the human race can visit upon itself is the act of war. However the very process also drives technological development forward. Witness the birth of powered aviation at the start of the 20th Century. In those early years the airframes were fragile, of stick and string construction, and the powerplants were, to say the least, unreliable. As the 1914-18 war progressed, the aircraft became stronger in construction while engines became more powerful and reliable and armament became part of the airframes' furniture.

The cessation of hostilities in 1918 brought the inevitable slowdown in technology development as nations regathered their strength to repair the damage caused by the war. On top of this, Germany was under a stronger burden trying to rebuild whilst paying reparations to the allies for damages caused. Under the Treaty of Versailles, the German armed forces were restricted in the types and sizes of weapons they could manufacture whilst the governments of the allies were redirecting their resources towards other projects. This left their air forces labouring with biplanes powered mainly by uncowled radial engines. Although these aircraft were the pinnacle of their type, the very design had reached the maximum performance limits possible.

Monoplanes were an obvious next stage in airframe development, although first efforts were no more than biplanes with their top mainplanes and associated struts and wires removed. Enter the art of developing aircraft for the purposes of racing. Point to point, course navigation, all served to drive the development of the small single seat, single-engined, high-powered racing

Prior to the appearance of monoplane fighters in the service of the RAF the primary frontline squadrons were equipped with biplanes. This is a preserved Gloster Gladiator wearing the prewar scheme applicable to most fighters of the era. (Nick Challoner)

plane. The story of the development of the Spitfire really begins at this point in the era of the seaplane races for the prestigious Schneider Trophy.

Prior to the start of the competition, Supermarine had achieved a measure of fame as a builder of patrol flying boats for the Royal Air Force (RAF). Whilst the boats were being designed and delivered, Reginald Mitchell was appointed the Supermarine Chief Designer in 1919 in recognition of his outstanding skills as a designer. The company's' first effort, the Sea Lion biplane flying boat, won the Schneider Trophy in 1922 against teams from Italy and the USA. Following on from this success came the S.4 monoplane float plane. This was a very streamlined design which managed to achieve a World Speed Record of 226 mph for floatplanes. Seen as a winner, the aircraft was then prepared to participate in the Trophy races. However, whilst undertaking high speed trials flights the mainplane developed an uncontrollable flutter which subsequently departed the airframe. Fortunately the pilot would survive the subsequent crash.

In 1927 Supermarine, in conjunction with the RAF High Speed Flight, tried again with the Schneider Trophy. In this attempt to win, a configuration similar to the earlier ill-fated S.4 was developed: the S.5. During the race, which it subsequently won, the aircraft achieved a maximum speed of 281 mph. Two years later Supermarine was to produce the definitive S.6 which won the trophy for the third time achieving a top speed of 357 mph in the process. A slightly developed version, the S.6B, not only won the Trophy for the final time in 1931, it also managed to set a further World Speed Record of 407 mph a year later.

Although high-speed float planes brought copious amounts of glory, few orders would result as the market was limited for such exotic aircraft. In fact, the total build for the racing series of floatplanes achieved a total of eight airframes. To earn their bread and butter, Supermarine continued the servicing and modification of the Southampton flying boats. However, this work was of limited value in providing capital for expansion. To remedy this situation, Vickers Engineering made a major investment in 1928 attaining a majority shareholding in the process.

Against this background of glamorous air racing the frontline Royal Air Force was still flying biplanes

The full effect of the streamlining principles applied to the S.6B are exemplified here in this side view. Like its predecessor, the Spitfire also appeared on floats although in lesser numbers than the Supermarine S series. (C P Russell Smith Collection)

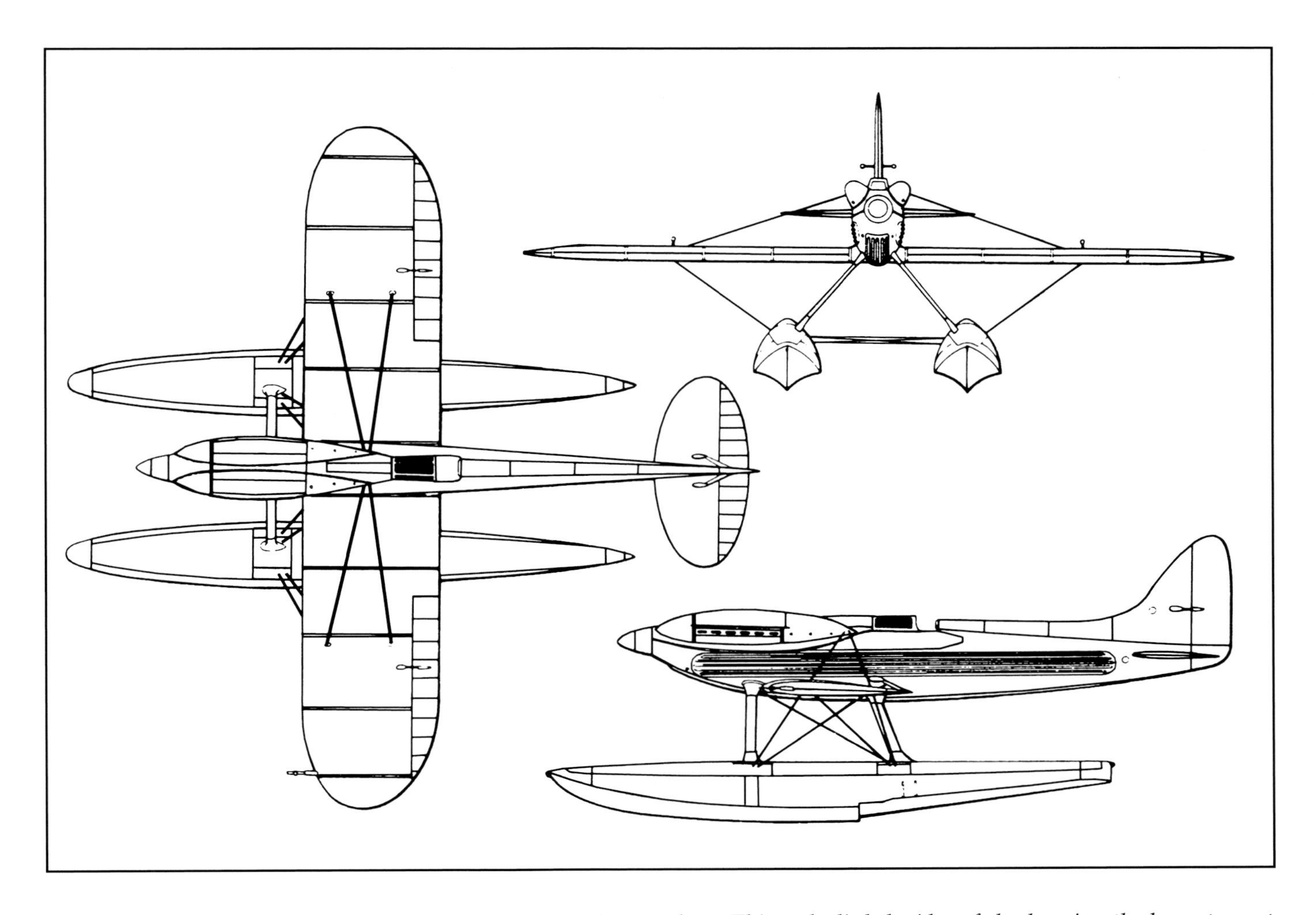

The Spitfire story starts with the Supermarine S.6B racing seaplane. This embodied the idea of shoehorning the largest, most powerful engine into the smallest, most streamlined airframe possible. (Big Bird Aviation Collection)

from the likes of Bristol, Gloster, and Hawkers. In contrast, the supposedly restrained German aircraft manufacturers were developing new aircraft for the Luftwaffe. One of the prime contenders on the fighter front was Messerschmitt with the Me-109, a highly-manoeuvrable, well-armed monoplane fighter.

Finally recognising that the frontline of the Royal Air Force was falling behind in fighter development, the Air Ministry issued Specification F.7/30 in 1931 in an attempt to rectify the deficiency. Primary requirements in the Specification document concentrated on six major points. These were defined thus: 1. Highest rate of climb possible; 2. Highest possible speed above 15,000 feet; 3. Best all-around vision possible for the pilot; 4. As manoeuvrable as possible; 5. Ease of manufacture for high-volume production; and 6. Ease of maintenance.

In response, the following companies sent in proposition documents to the Air Ministry: Blackburn, Parnell, and Westland all sent biplanes proposals; whilst monoplanes came from both Boulton and Paul Aircraft and Supermarine. After initial evaluation, the Air Ministry chose the projected Supermarine Type 224 for further development, issuing a contract for that very purpose in late 1932. Having chosen Supermarine to develop the new fighter further, definitive service requirements were issued. These included wing-mounted machine guns and the ability to tote four 20 lbs bombs for ground-attack purposes.

The resulting aircraft that rolled out in early 1934 was radical in design in many respects. One radical change was the unusual cranked mainplanes which did away with the need for external bracing, a feature of earlier monoplanes. However, the design did reveal some caution in that the wing thickness was deeper than it needed to be. The reasoning behind this was based upon eliminating the possible wing flutter problems that had befallen the earlier S.4. Included in the wing leading edges were radi-

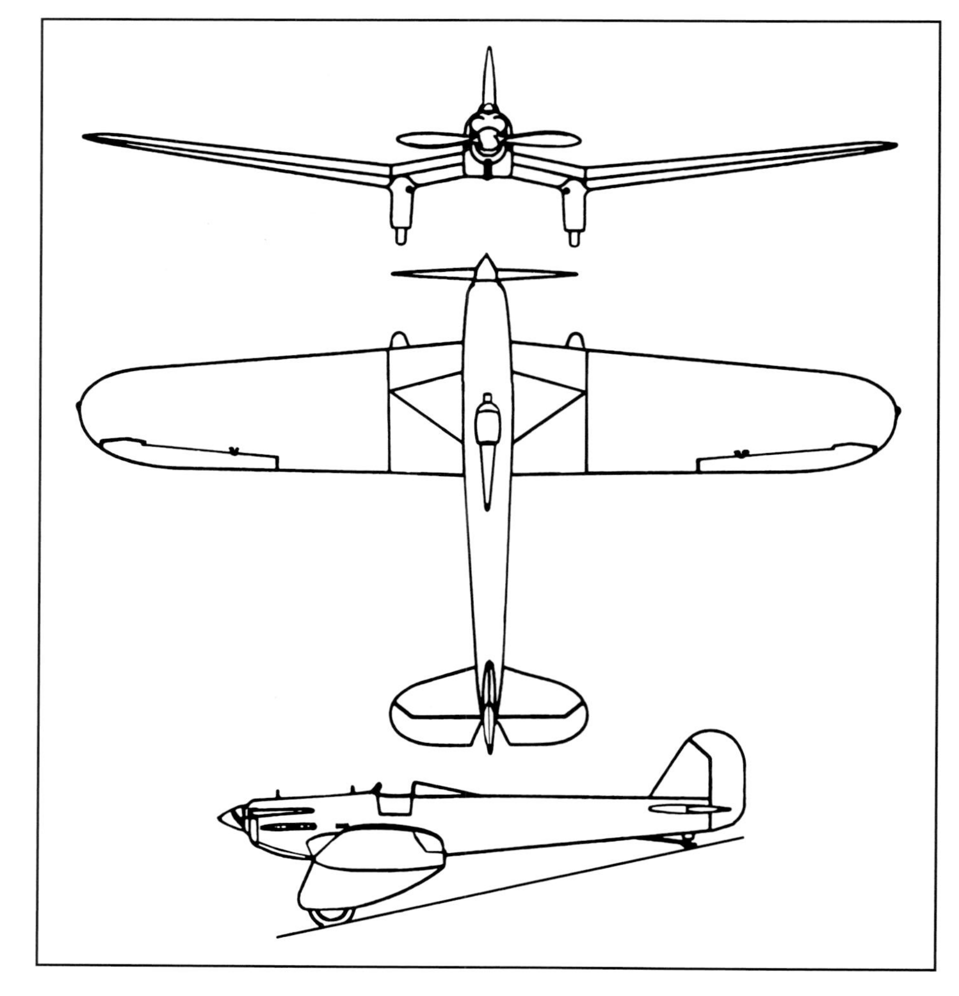

The first attempt by Supermarine to produce a fighter resulted in the crank-winged Type 224. However the cooling system for the Goshawk caused severe performance problems. (Big Bird Aviation Collection)

The next stage in the development of the Spitfire was the Supermarine Type 224. Hangovers from an earlier period of aircraft design are the trousered undercarriage legs and fixed-pitch propeller, whilst the innovations are the cranked wings within which are buried the radiators. (Eric B Morgan Collection)

ators and condensers for cooling the Rolls-Royce Goshawk engine rated at 660 hp, which was the most powerful unit available at the time. The coolant for this unusual installation was housed in tanks installed within the trouser fairings for the main undercarriage units.

First flight of this Supermarine's fighter was undertaken during February 1934 and immediately problems with the engine cooling system arose. Normally an engine of this kind had its coolant system concentrated in the upper wing of a biplane which then used gravity to feed the condensed steam as water back to the engine coolant system. However, the design of the Type 224 meant that the condensers were fitted in the undercarriage fairings and relied upon pumps to push the steam back to the header tank above the engine. As the system required that coolant be fluid to achieve its purpose, the return of steam to the engine would cause overheating. This was the main problem that befell the new Supermarine fighter from the outset. Flight after flight the red overheat warning lights came on causing flights to be aborted. Even when the Type 224 did manage to carry out some flight testing it suffered further problems when climbing to height. To counter this the only recourse open to the pilot was that of leveling out to allow the system to cool down.

All of these problems could have been resolved had the Type 224 not suffered another even more major problem. However, its performance was abysmal especially in comparison with the successes achieved by the company's floatplane racers. Top speed was pegged at 238 mph at an altitude of 15,000 feet and the rate of climb to the same height took a long eight minutes according to one pilot.

The final contest winner was eventually declared as the Gloster SS.37 which entered service with the RAF as the Gladiator. Although the Type 224 was very much seen as a failure, it did undergo some testing at Martlesham Heath before passing onto Farnborough for further flight testing. During this time the aircraft was temporarily named *Spitfire*. Eventually the prototype ended its days as a target at the Orfordness firing range.

Even as the Type 224 was not proving its worth as a contender for the vacant RAF fighter slot, Supermarine was in consultation with senior officials at the Air Ministry where they proposed a fighter design that owed more to the racing floatplanes than the 1931 specification. The new proposal was a very refined aircraft compared to the Type 224. Designated the Type 300 the new aircraft had lost the corrugations in the wing leading edge which was also straightened out. A reduction in span by six feet and the installation of a retractable undercarriage saw the designers offering an aircraft that was capable of reaching 300 mph using the Goshawk engine. By the end of July 1934, the team at Supermarine offered the reviewed aircraft to the Air Ministry as Specification 425a.

Although this next manifestation towards the Spitfire looked correct to the designers, the officials at the Air Ministry were less than impressed as the projected performance given by the Goshawk engine was no greater than that of other fighters available at the time. Undeterred, Mitchell and the Supermarine design team began to revise the arrangement of the aircraft whilst looking at an alternative powerplant as the evaporative cooling system of the Goshawk was proving troublesome.

Initially this was to be the Napier Dagger which had a projected rating in the future of 800 rpm. However Vickers and the Supermarine board decided against this engine in favour of a new powerplant being developed by Rolls Royce designated the PV XII. The decision in favour of the PV XII was ratified in November 1934, soon after the name Merlin was bestowed

The contemporary fighter to the Spitfire was the Hawker Hurricane. Also powered by the Merlin engine the Hurricane's antecedents meant that it was slower than the Spitfire. This fact notwithstanding, the Hurricane actually shot down more enemy aircraft than any other British fighter. (Nick Challoner)

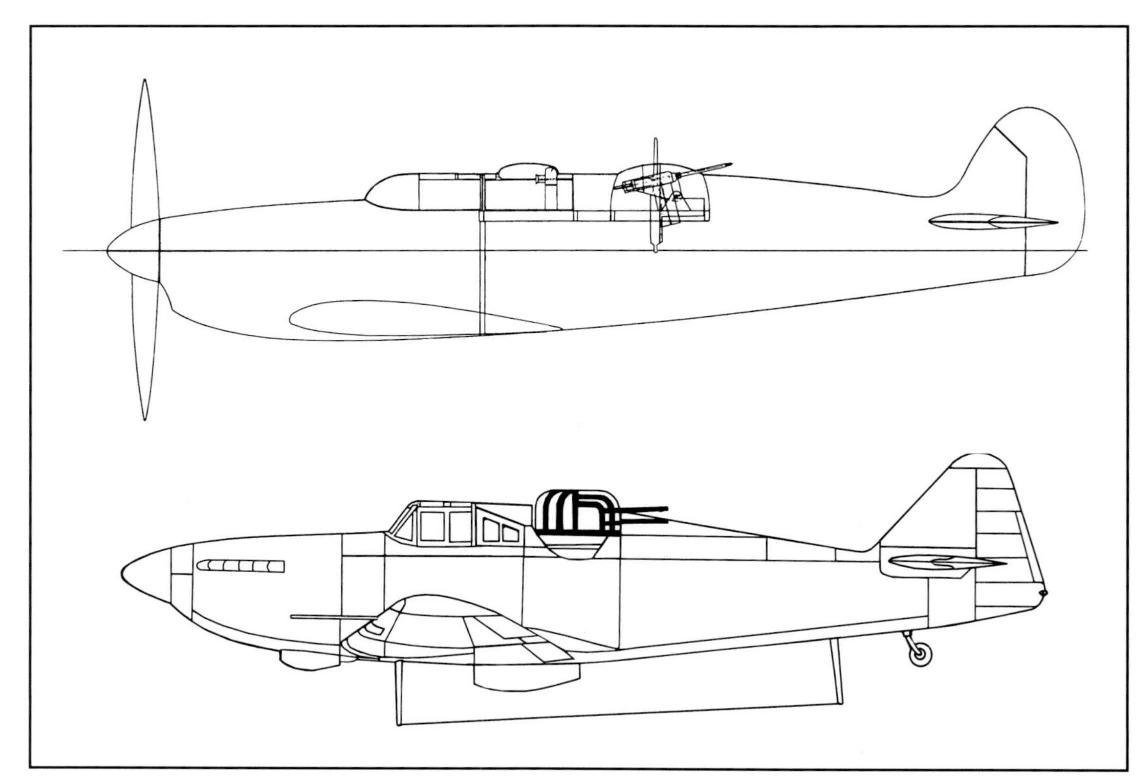

One of the side paths of design followed by the Supermarine team was that of the turret fighter as shown in the upper part of this view. The lower side view shows the Boulton Paul Defiant which eventually won the competition. However, the Defiant was not very successful and most eventually ended up involved in second line duties. (Big Bird Aviation Collection)

upon the new engine. Ground running of the engine had already achieved a rating of 625 hp, which was increased to 790 hp at 12,000 feet. Although the new powerplant was experiencing the usual teething troubles, it was projected that the production versions of the 27 litre engine would be able to generate 1,000 horsepower at maximum output.

Having found the right engine for the airframe the Air Ministry was approached with a revised design by Mitchell and Supermarine. In response the Ministry issued a contract, AM. 361140/34, on 1 December 1934 for one prototype to the revised F.7/30 design which was valued at £10,000. First flight for the new aircraft was set for October 1935. Very few changes were made to the actual specification which still retained the evaporative system in the wings, although the armament was revised to include four machine guns. Revision of the actual design was undertaken by the Supermarine team which had to redefine the airframe to take account of the increased weight of the Merlin.

The area affected by the changes was confined to the wing where the leading edge underwent some revision. It was only a short step to the tapered elliptical wing that was to be such a feature of the prototype and production aircraft. The selection of an ellipse shape was also seen as a perfect aerodynamic compromise for a highly manoeuvrable fighter combining the best coefficients of drag and lift. The decision to include a pronounced taper also allowed for a retractable undercarriage to be provided from the beginning.

The wing design was unique in other ways. Not obviously visible to the casual viewer is the gentle twist in the wing which started at the root and extended all the way to the tip. Such a feature ensured that the wing root would stall well before the tip, this meant that pilots had better warning than most of an impending stall. Construction was also unusual in that it was intended to combine both strength and lightness. The core for this was the main spar which was constructed of square sections of different sizes placed one inside the other. A web was then placed between two of the tapered booms to which was attached the leading edge ribs. This assembly was then covered by a heavy gauge skin which provided a strong "D" shaped torsion box that was to feature throughout the life of the Spitfire.

On the rear of the spar at the root was mounted the main undercarriage unit. Upon retraction, this unit moved outwards and slightly to the rear and sat quite happily in the nonload bearing area of the wing. Although the undercarriage presented a very narrow track this was seen as acceptable as it reduced the loads generated by landing upon the wing structure. Initially the evaporative system for the engine featured an intake leading to an auxiliary radiator situated behind the cockpit. This would be opened during the take off and climb phases, but would be shut in cruising or high-speed flight mode. Air intakes for the carburetor were to be located in the wing roots whilst the oil cooler intake was to be situated under the nose leading to a pair of ducted outlets under the fuselage. As the new aircraft was also intended for the night flying role, great consideration was given to the placement of the engine exhausts. Initial thoughts led to the development of exhaust ducting that vented out under the fuselage.

As the evolution of the Type 300 continued further thoughts turned to the aircraft's armament. The initial four machine guns were seen as insufficient in view of the all-metal bombers entering service with many air forces in Europe. An updated version of the specification, 10/35, revised the wing-mounted gun requirements up to eight Browning machine guns. Further design work encompassed the placement of the tailplane and the location of the rudder hinge line. Initially the tailplane had been located in the mid fuselage position, although it was soon moved further up to the top of the fuselage as the earlier point had shown spin recovery problems on a 1/24 scale model built to cover such eventualities.

Concern about the method used to cool the engine was still giving cause for concern especially in view of the heat generated by the PV XII. It was thought that the evaporative system could end up being the downfall of the emerging fighter. Therefore, another method was required even though Mitchell had expressed a desire to eliminate any external intakes. The intake chosen to cool the engine was designed by Frederick Meredith whose ducted radiator system was fed through a cooling matrix which in turn compressed the air and expelled the exhaust. This in turn would generate enough thrust to overcome any drag. In order to improve the performance of the cooling system it was decided to use ethylene glycol as the coolant medium, it was also found that such a system was far lighter than the water equivalent.

When the air forces of Germany and Britain finally came head-to-head, the Luftwaffe pilots were mounted upon the Messerschmitt Me-109, which had been tried and tested under combat conditions in the Spanish Civil War. (Big Bird Aviation Collection)

The evolution of the Spitfire was accelerated by the events that took place after September 1939, thus the Mk.V became one of the most produced fighters in the early years of the war. This preserved example wears a desert camouflage scheme and features the prominent strengthening angles above the undercarriage bays. Problems in this area of the wing were to plague the Spitfire throughout its service life, although stress problems were reduced by clipping the wings. (Dave Stewart)

Whilst the basics of the design were being hammered into shape the drawing office was looking at other ways of extending the design. First on the drawing board was to be a two-seat fighter, the second seat being occupied by an air gunner operating a four-gun turret. This aspect of development was later dropped upon the emergence of the Boulton Paul Defiant fighter which was less than successful in its intended role. The second project covered a two-seat trainer as it was acknowledged that the emerging fighter, being a thoroughbred, would be more than a handful for tyro pilots. Although the trainer was seen as a necessity it took until the emergence of the Spitfire Mk.IX for there to be any positive steps in that direction.

Building of the first prototype had begun at the Woolston factory in late 1935, work starting first on the wings. As the fuselage began to be built some changes were incorporated into the design. The monocoque fuselage aft of the cockpit was changed to an oval shape, this in turn allowed extra glazing to be included aft of the sliding canopy, which in turn increased the pilot's vision area. Forward of the main cabin bulkhead was mounted an early example of the Rolls-Royce Merlin engine which in turn drove a wooden two-bladed fixed pitch propeller. The primary cooling radiator matrix was finally located in its production position under the starboard wing aft of the undercarriage bay whilst the oil cooler venturi occupied a similar position under the other wing. Overall finish of the airframe was far in advance of the norm for the aviation industry at the time as it was mainly a flush-riveted assembly to reduce the drag of the airframe.

Although building of the prototype was proceeding apace there was some concern raised by the Air Ministry about its first flight date, especially in view of the rearmament of the German armed forces and the Luftwaffe in particular. Britain's answer to this change in the balance of power in Europe was to start an emergency rearmament programme, part of which required the building of 900 fighters. This total was divided into 600 airframes from Hawkers, in the shape of the Hurricane, whilst the remainder were to be built by Supermarine in the shape of the Spitfire.

The initial proposal from Vickers was to deliver the fighters on time if the order for such was placed at an early date. In order for this to work the manufacturer's trials would be completed by the end of March 1936 with trials at Martlesham to be completed a month later. The order would then need to be placed in the following month for production deliveries to commence in September 1937. From this point the programme began to slip behind schedule. Instead of flying at the beginning of 1936 as expected, ground tests did not commence until February. Successful completion of these trials meant that the prototype, later named Spitfire in May 1936, by now given the Air Ministry serial K5054, would be ready for its maiden flight at the beginning of March.

On the 6th of that month the Supermarine company's chief test pilot, Mutt Summers, lifted the pale green prototype F.37/34 into the air and a legend was born.

SPITFIRE GENESIS

2

THE BIRTH OF A LEGEND

The first flight of K5054 was to last no longer than twenty minutes and reached an altitude no higher than 5,000 feet. During the aircraft's brief sojourn in its natural element the test pilot, Mutt Summers, assessed that the technical team has ascertained the correct stalling speed after which a simulated landing was carried out. With initial handling trials carried out, the aircraft was turned for home and landed successfully. The post flight debrief from the pilot revealed that no major faults had been encountered during this first airborne venture and that he would be happy to fly it again as it stood.

There were some changes made to the prototype prior to its second flight, although these were planned upgrades rather than rectification. The original fine pitch, two-bladed propeller was replaced with one of a coarse pitch which would allow the new fighter to reach its maximum computed speed. During the first test flight K5054 had been flown with its undercarriage locked in the down position. For the second flight the gear was unlocked and the fairings fitted after which a further set of functionals was carried out to ensure that clearances and operational behaviour were correct.

Cleared for flight, K5054 was next flown on 10 March during which the undercarriage was cycled through its range without any problems. A further flight to consolidate the results gained on the earlier flights was carried out the following day. A report submitted a few days later to the Air Ministry stated that initial test flying was successful and that all systems including flaps and undercarriage appeared to be working normally. It also stated that although the aircraft had not been flown to its maximum speed, its potential should see it quite successfully exceed the Hurricane's top speed. The only complaint from the pilot was that there was, and always would be a problem with forward and downward vision.

A further series of test flights followed before the prototype returned to the hangar for a much needed engine change. During the test flight series there had been a problem with intermittent engine starting which had delayed some of the sorties. Fortunately, Rolls-Royce had a spare unit available which was immediately shipped to the hangar at Eastleigh. With its power unit replaced the prototype resumed test flying, this time the intention was to push it to its maximum speed possible. This was achieved on 25 March when K5054 was climbed to 17,000 ft. from which it was entered into a shallow dive. During the 50 minutes of the flight the aircraft achieved an airspeed of 350 mph indicated which was equivalent to a true air speed of 430 mph.

Although much of the glamorous test flying had been completed, a further series of flights was required to establish the correct parameters for the airspeed indication offset errors. In order to calibrate the graphs and aircraft correctly, all the flying was done in the early hours of the morning as this was the only period of the day guaranteed to produce the still air conditions required. A sequence of

Jeffrey Quill, Supermarine chief test pilot at the controls of the prototype Spitfire, K5054, prepares for an early test flight. Still designated the Type 300 this, the first of the breed, exhibits many differences from the production version. These include suppressed exhausts and a tail skid instead of a tail wheel. (Eric B Morgan Collection)

evaluation sorties over a variety of speed ranges eventually established the correct parameters. As the flying progressed and operational speeds increased, the test pilots began to notice that there was a handling problem with the rudder which was found to be light in the upper-speed ranges. Investigation revealed that the rudder horn balance was too large which made the aircraft longitudinally unstable. Corrective action was also required to enable the aircraft to reach the predicted speed of 350 mph in level flight.

To achieve this K5054 was grounded in early April and moved into the hangar at Eastleigh for ground resonance testing. Once completed the modification team set about incorporating tweaks and changes to the airframe and engine. The first area looked at was the rudder which had its horn balance reduced in size and its top flattened. Remedial work was also required to strengthen the engine cowling panels which had been reported as rattling in flight. Another area requiring some attention was the carburetor intake which was lowered slightly to increase the ram air pressure. Once completed the aircraft was prepared for painting. To achieve the finest finish possible Supermarine called in the finishing specialists that painted the car bodies for Rolls-Royce. After hours of filling and rubbing down the primer coats, the final blue grey paint cover was applied.

The revamped aircraft was ready to resume test flying at the beginning of May and it was at this time that a name was finally bestowed upon the Type 300: Spitfire. It was reported that Mitchell was not too impressed with the name, but it was accepted anyway. On 11 May the repainted and named fighter prototype resumed its intensive test flight schedule. The first result of this series of flights was that the Spitfire was found to have better handling in the air courtesy of the revamped rudder. As the control problem was no longer bothering the aircraft, further flights concentrated upon reaching the maximum speed possible.

It had already been deduced that the wings fitted to K5054 would limit its ability to achieve 350 mph in level flight, however such was the pressure exerted upon those participating in the development programme that it was decided that the prototype would continue flying unchanged. Therefore, to achieve the speeds necessary, all high-speed runs would be carried out in a shal-

Seen from a slightly different angle the Spitfire prototype reveals other traits that were peculiar to this aircraft. The canopy hood is of the earlier flattened variety that was quickly replaced in service by a more bulged assembly which allowed the pilot greater all-round visibility. The extended pitot head on the port wing was later replaced by a smaller version under the wing. However the installation of such an item was necessary for calibration of the aircraft's speed envelope. (C P Russell Smith Collection)

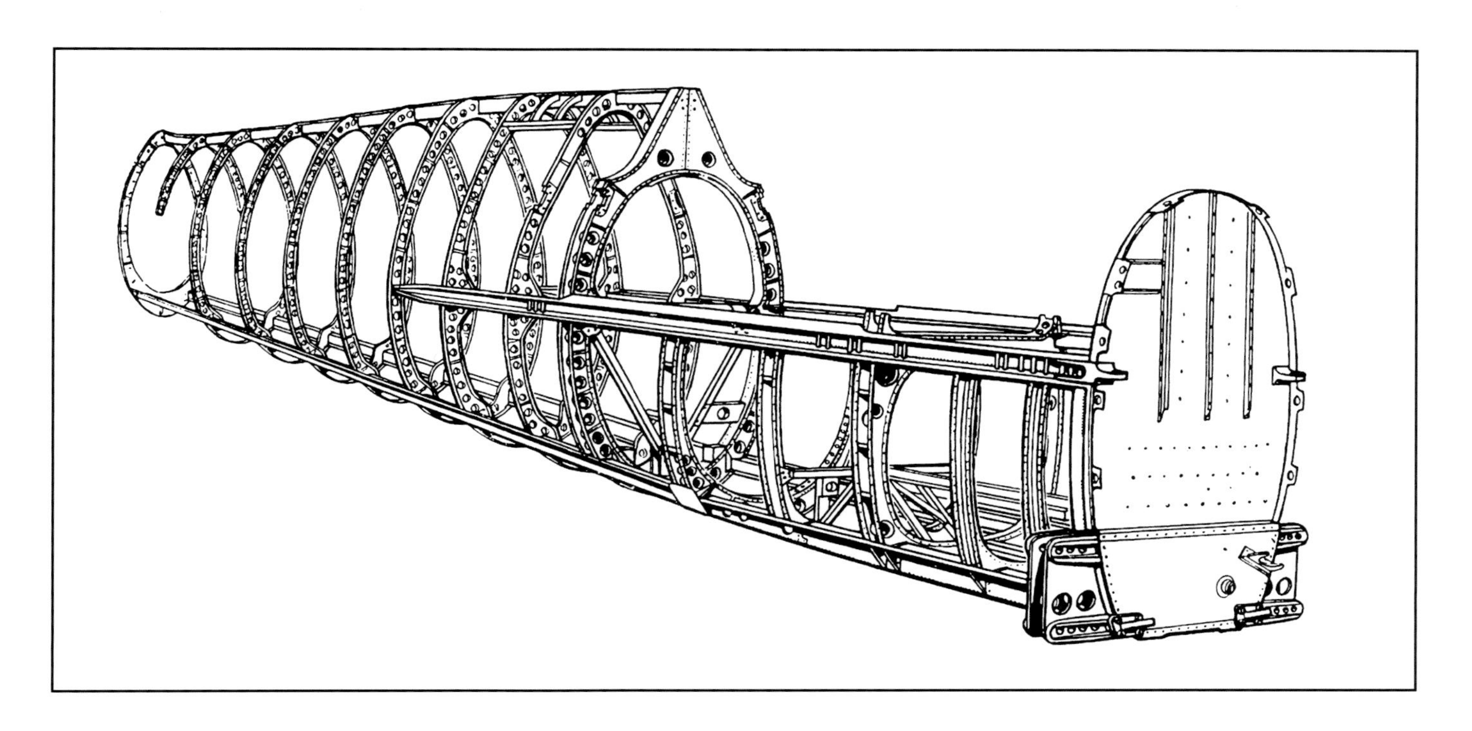

This structural view of the Spitfire fuselage starts at Frame 5 and proceeds aft. Throughout its production history this layout remained virtually unchanged except for changes forced through modification action. (Chris Michell)

low dive. During a series of these controlled dives the Spitfire prototype managed to achieve a maximum speed of 380 mph accompanied by a large bang. A cautious return to base was accomplished where it was discovered that one of the undercarriage fairings had broken clear and hit the fuselage as it departed. Fortunately the damage was assessed as minimal.

Once the damage had been repaired the Air Ministry began to put pressure on Supermarine to deliver the prototype to Martlesham Heath, the forerunner of A&AEE Boscombe Down, for evaluation flying. The Hurricane from Hawkers was already undergoing evaluation and it was realised that if the Spitfire were to arrive too late the chances of the production contract forthcoming would be reduced if not canceled. However, the technical team wanted to give K5054 the extra speed advantage that had been promised from the outset. Investigations into the factors that could be limiting the aircraft's performance centred upon the propeller and its tip speed which was found to be entering high Mach speeds. To counteract this a new propeller was manufactured which

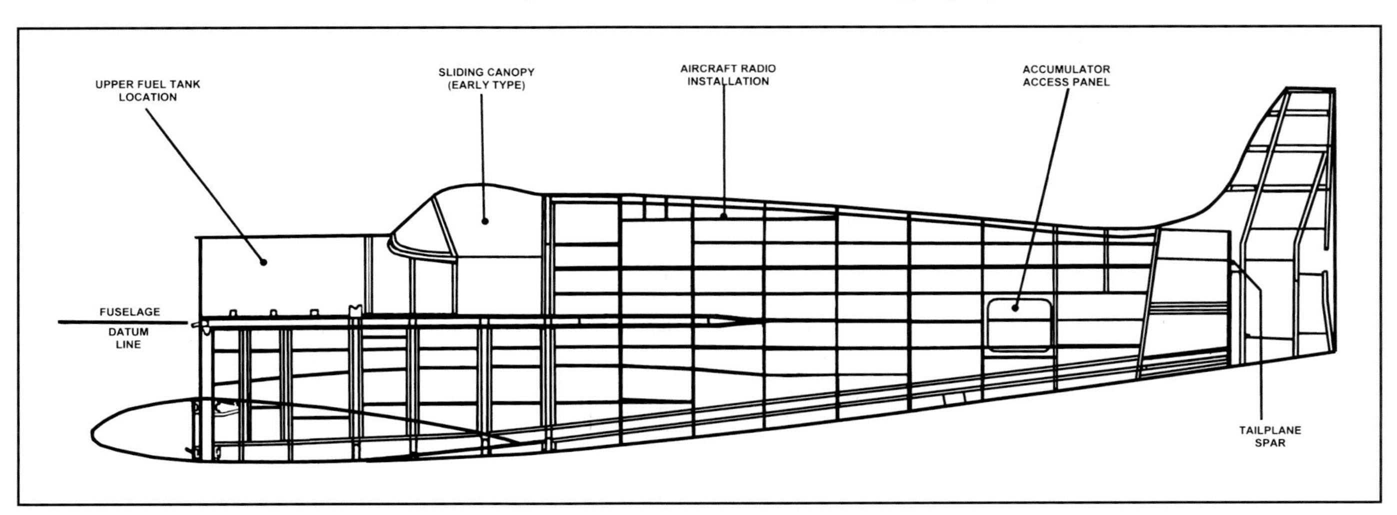

The structural layout and position of the mainframes remained unchanged throughout the life of the Spitfire. (Chris Michell)

reduced the high speeds at the tips. First test flying of the unit gave Supermarine the increase in performance that had been hoped for. During this flight a maximum speed of 348 mph was achieved which gave the Spitfire an edge of 13 mph over the Hurricane.

With company test flying completed the Spitfire was delivered to Martlesham Heath on 26 May, although it was minus armament as there was a lack of Browning machine guns available. Thus ballast weights were installed in the wings of the Spitfire whilst the Hurricane had the available weapons. Unusual for a new fighter, test flying by both service and evaluation pilots began almost immediately. After completion of the first flight the Air Ministry was informed that the aircraft was more than adequate for its task. On the strength of this report and the subsequent sequence of flights, the Air Ministry signed a contract with Supermarine for 310 aircraft to be known as the Spitfire Mk.I on 3 June.

The heart of the Spitfire wing was the multi-part tubular assembly main spar. This is one of new manufacture awaiting final assembly by staff at Airframe Assemblies. (Chris Michell)

Although the Ministry had ordered production Spitfires, further tests and trials were required on K5054. The first of these tests was an accurate weighing which required all fluids to be removed to leave the aircraft in its basic condition. Once the tare weight had been calculated, all the extra equipment from service sources such as the radio, guns, and ammunition, all represented by ballast, were fitted. A full-fuel load plus oil for the engine brought the airframe up to its full service operating gross weight which was calculated at 5,359 pounds. Further measurements established the C of G aft limit at 9.9 inches aft of the datum whilst the forward was located some 8.4 inches from the same point.

Throughout the remainder of June the prototype was pushed to the limit and finally managed to achieve a top speed of 349 mph at an altitude of 18,000 feet, the powerplant was still the original 900 horsepower Merlin Type C. However, not everything was perfect with the Spitfire as the final report shows. Some of the items mentioned included the statement that the windscreen would need redesigning as the current fitted item was restrictive of vision whilst in flight. Other minor problems mentioned included the undercarriage warning Klaxon which was found to be too quiet when the canopy was open, the elevator trim would require regearing as it was too sensitive, the radiator flap was difficult to close at high speeds and that a rethink of the location of access to the radio for servicing would be required. These minor defects notwithstanding the Spitfire was praised as a highly manoeuvrable aircraft, with performance and handling that would only improve when the production version of the engine was installed.

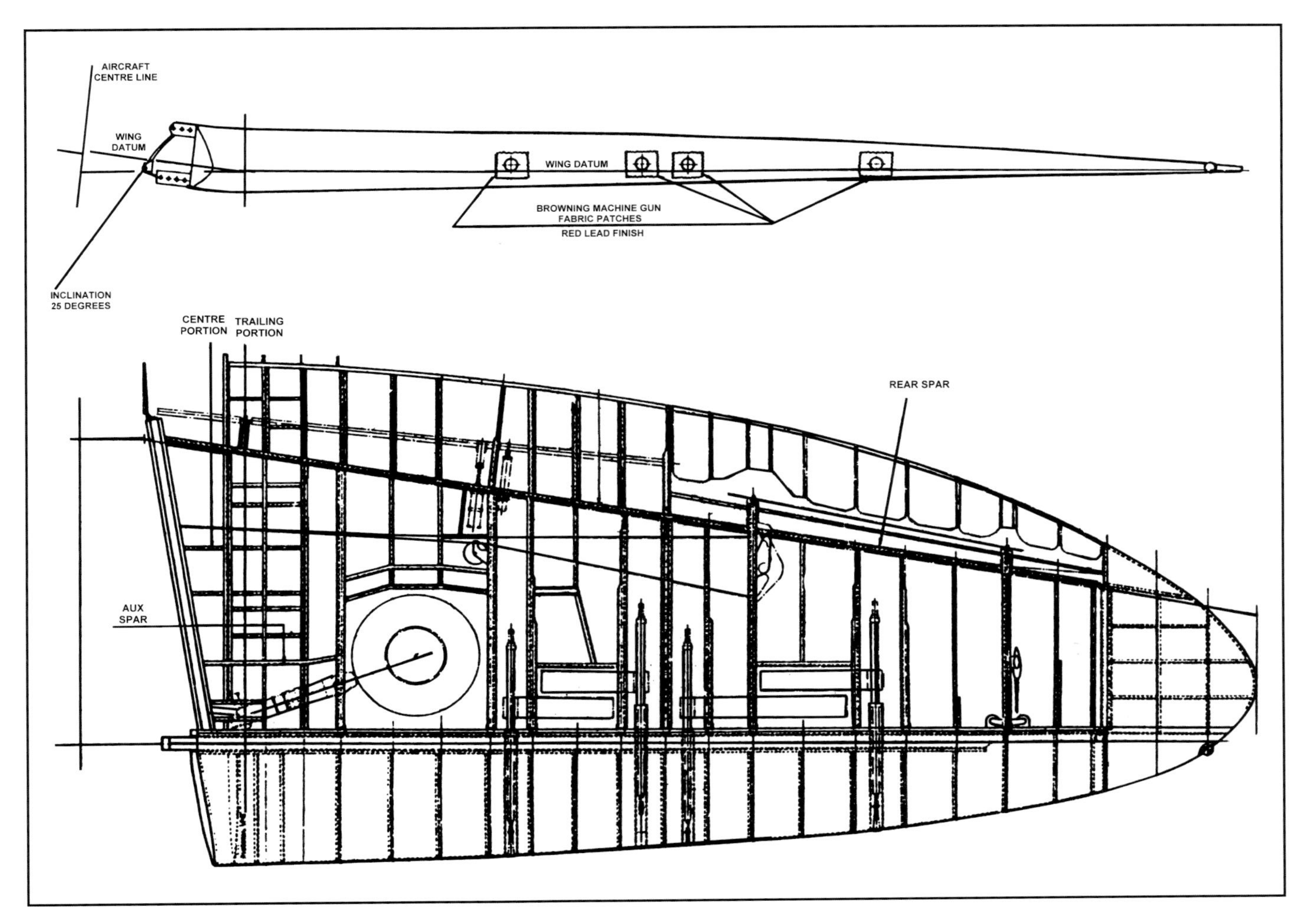

Changes in role and the final emergence of the Griffon-powered aircraft, notwithstanding the layout of the Spitfire wing structure, retained a similarity to the prototype during the entire production run. (Chris Michell)

On 27 June the prototype was flown to Hendon for the Royal Air Force Pageant from whence it departed for the SBAC show at Hatfield some two days later. Having appeared to an enthusiastic public K5054 returned to its assigned testing duties. The primary requirements covered at Martlesham Heath in these resumed flights concerned, as before, height and speed trials. On an early flight in July the aircraft was flown to its highest altitude so far of 34,700 feet. Not long after this increase in the Spitfire's perfor-

Seen standing in a storage rack are a pair of Spitfire wings awaiting fitment to a completed fuselage. Clearly visible is the flap and the associated driveshaft plus the bearing hinges for the aileron. The main gear leg is shown retracted whilst the structure that helps support the oil cooler is clearly visible. (Chris Michell)

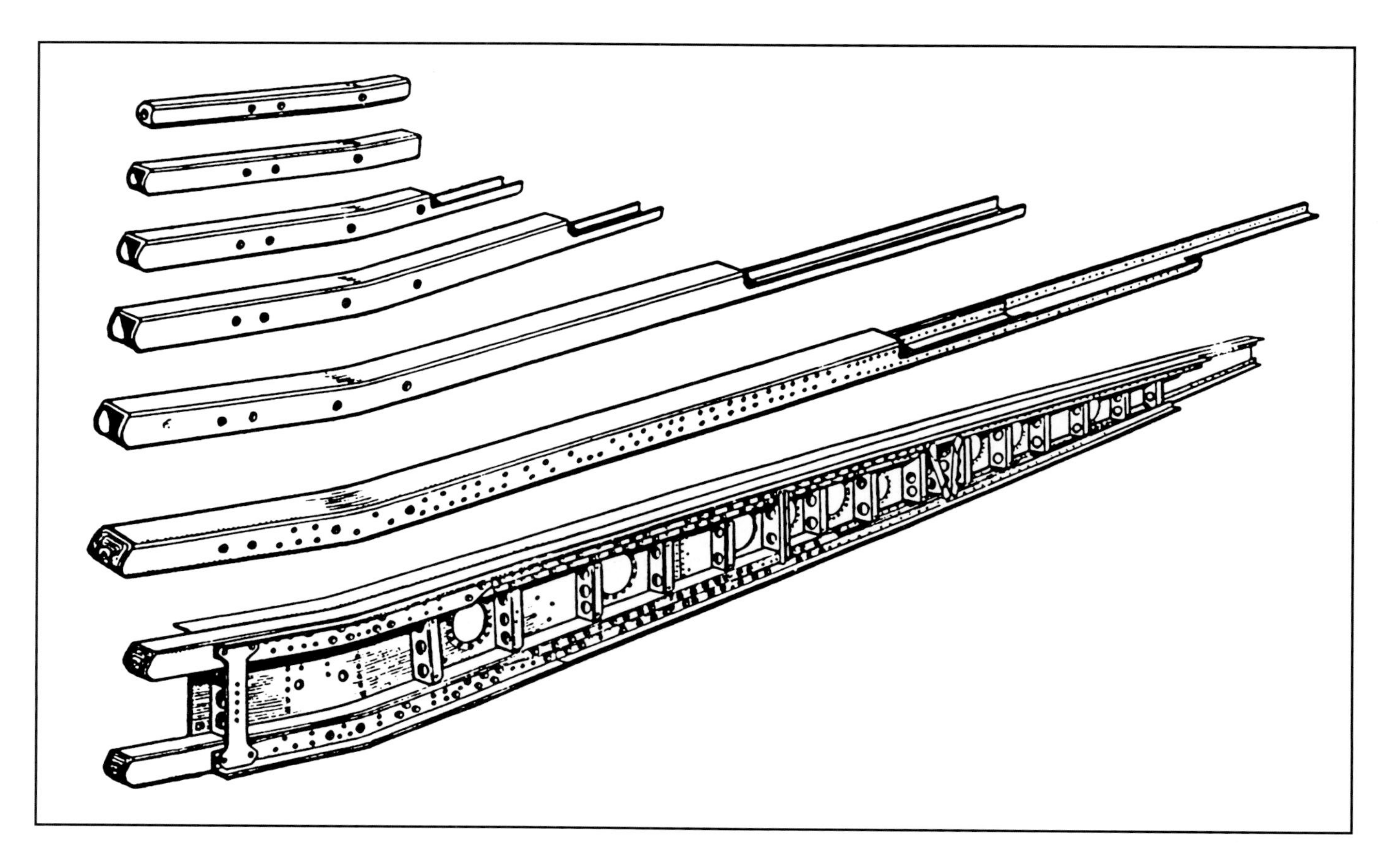

This diagram fully illustrates the complicated nature of the wing spar shown here in unassembled form. (Chris Michell)

mance envelope was achieved, a small amount of modification work was carried out on the undercarriage legs. As built, each leg had a sprung fairing at the top and bottom of the fixed portion the purpose of which was to present a clean surface to the airflow. However, on some flights not all the hinged portions behaved as expected as on occasion the odd item remained out in the airflow. Although not highly detrimental to the Spitfires performance it was decided to remove them completely. Comparative test flying discovered that there was very little difference in handling with or without the fairings fitted.

With the greater majority of flight testing completed it was now the turn of the armament to be evaluated. For this purpose K4054 was returned to Eastleigh for the installation of the eight Browning machine guns and associated ammunition tanks. During September the gun and tank installation was completed as was that of the reflector gunsight. Also accomplished at the same time was the fitting of the radio. During the following month an uprated Merlin F rated at 1,050 horsepower was installed in the prototype as was an improved oil cooler and undercarriage jacks. As the prototype still had to undergo spinning trials, an anti-spin chute and fairing was also installed.

Test flying of the aircraft resumed on 3 December 1936 and mainly concentrated upon the performance of the new engine and oil cooler. Following on from these flights the Spitfire was test flown with a variety of propellers although as they were of fixed pitch no great improvements could be discerned. Once these were completed, the original assembly was refitted and the Spitfire prepared for spinning trials. On 7 December the aircraft was spun for the first time successfully with the aircraft trimmed towards the fully forward position after which the same series of flights was carried out to cover the fully aft setting.

Flying of K5054 resumed early in the new year of 1937 and concentrated upon the "split pea" trials. As built the prototype was fully flush riveted throughout, however this method of assembly was slower and more expensive that that using dome head rivets. To represent the dome heads, split peas were glued to various parts of the airframe to see what deterioration, if any, there was in performance in handling. Although crude by modern standards the trials did reveal that under no circumstances should such fasteners be used on the wings as they were detrimental to the aircraft's performance. In contrast they could be used without too much deterioration in horizontal rows along the fuselage.

Another modification that featured on K5054 early in 1937 was the fitment of a tail wheel in place of the original streamlined tail skid originally installed. Although the tail wheel assembly would slow down the Spitfire marginally, the Air Ministry explained that another form of tail support would be needed once the hard standings, taxiways, and other infrastructure requirements had been built into the frontline air bases. Both single and double wheel installations were tested, the former being the one chosen. Having settled the tail wheel problem the Spitfire was returned to Martlesham Heath for gun firing trials. First firings were carried out using the ground butts and were successful, however airborne firing trials did not proceed so smoothly as only three of the guns fired. It was later deduced that the breeches had frozen in flight. Unfortunately upon landing, some of the breeches became unstuck and launched a round each into the air. Remedial work would later feature gun heating and modification to the breech safety mechanisms.

Although the prototype had led a charmed life up to this point, this ceased on 22 March when the aircraft suffered an engine failure. The pilot made an emergency wheels-up landing which caused only minimal damage to the airframe. Investigation revealed that there had been a bigend and connecting rod failure which had punctured the crankcase. Further investigation by Rolls-Royce showed that there had been a problem with the oil feed to the main bearings. Having been returned to Eastleigh for repair the opportunity was taken to install the gun heating system and to repaint the airframe in a more warlike earth and green finish. Once repaired and refinished the prototype returned to flying status during which time it was used to test a succession of propeller types. Once these were completed, trials began with ejector exhausts which were found to give extra thrust and a resultant increase in top speed to 360 mph.

Test flying continued on into the early months of 1938 and included night flying and gun firing trials with the heating system installed. At about this time service pilots began

The normal fighter versions of the Spitfire mainly sported two types of wing tips. This is the standard tip much favoured for medium- to high-altitude air combat. (Chris Michell)

In contrast to the standard wing tip many Spitfires were fitted with the clipped version for operations at low level which improved the type's rate of roll. Also note the pitot head under the wing. (Chris Michell)

MKVB
CLIMB
DESCENT
M.P.H.
REMOTE CONTACTOR
TYPE 4
FLOOD LIGHTS
NO SMOKING
ON
OFF

Opposite Page:

The opening of the cockpit door has allowed the photographer to show the pilot's control panel to great effect. Note the gunsight mounted on the coaming. (Chris Michell)

flying the aircraft, although not without mishap when minor damage occurred on a nose over after a series of landings. A further more serious accident was suffered by K5054 after a night flying sortie on 23 March when after a bouncy landing the port undercarriage leg was pushed up through the upper wing skin.

After returning again to Eastleigh courtesy of road transport for repair, K5054 resumed its flying career on 19 July complete with further modifications to the gun heating system. During the following months the prototype was flown at both Martlesham Heath and Eastleigh which concentrated upon curing the gun heating problem. By the end of October 1938 the problem was solved and the trials of K5054 came to an end.

However, this was not the end of the prototype's flying career as it was delivered to Farnborough in early November for trials in connection with the high-speed Spitfire with which it was planned to gain the World Speed Record. Both the pilots involved used the prototype as a type trainer before the Speed Spitfire went for the record. From its use as a conversion trainer, the prototype began a series of trials involving fuels of different octane's plus spark plugs and carburetor settings. The purpose of these flights was to ascertain the best performance possible for a high-speed fighter. Maximum speed achieved by K5054 was 380 mph.

This view of the cockpit encapsulates the airscrew control levers plus numerous gauges that indicate such services as oxygen contents, brake pressures, and the airspeed indicator. (Chris Michell)

The flying career of K5054 came to an abrupt end on 4 September 1939, the day after Britain declared war on Germany, whilst the aircraft was returning to base at Farnborough after testing a new type of spark plug. On approach to land, the Spitfire touched down far too fast even-

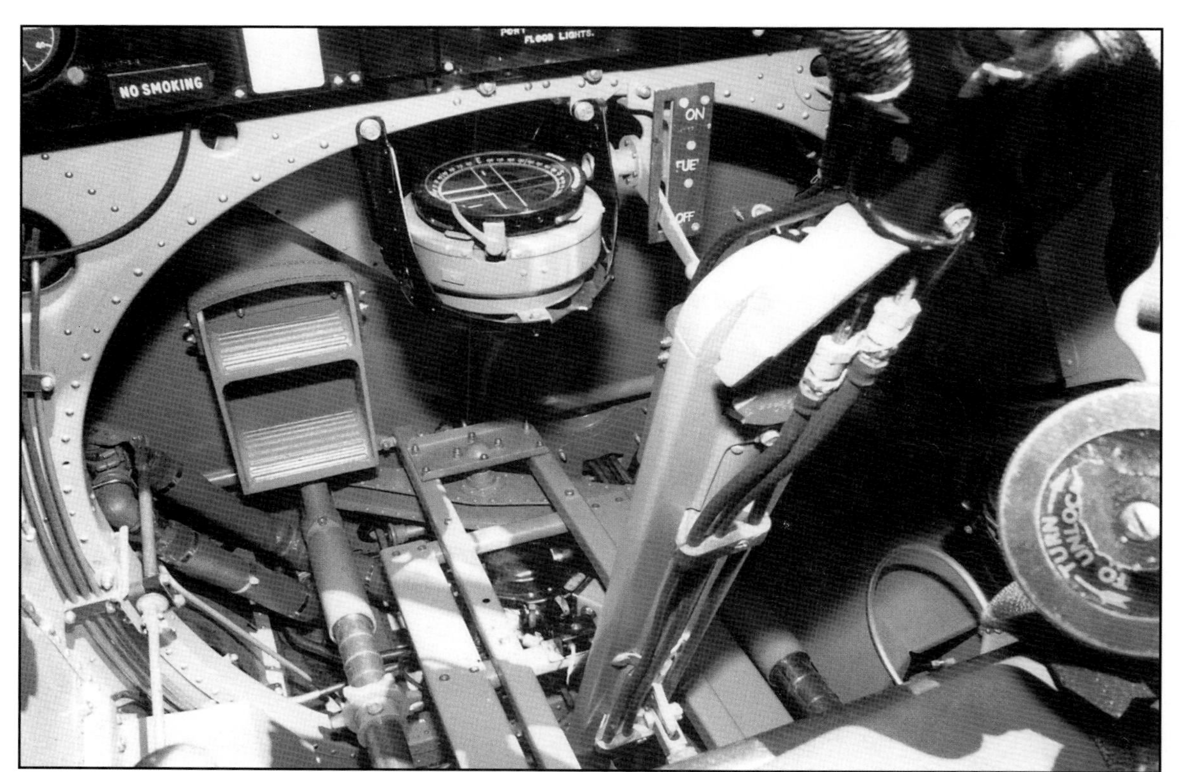

Very much under the dash shows the port rudder pedal and its guide plus the compass and mounting structure for the control column. (Chris Michell)

A slightly different perspective of the port side of the cockpit shows that famous control column, the compass, and elevator trim wheel. (Chris Michell)

Looking down the starboard cockpit wall reveals the undercarriage operating controls of the Spitfire. (Chris Michell)

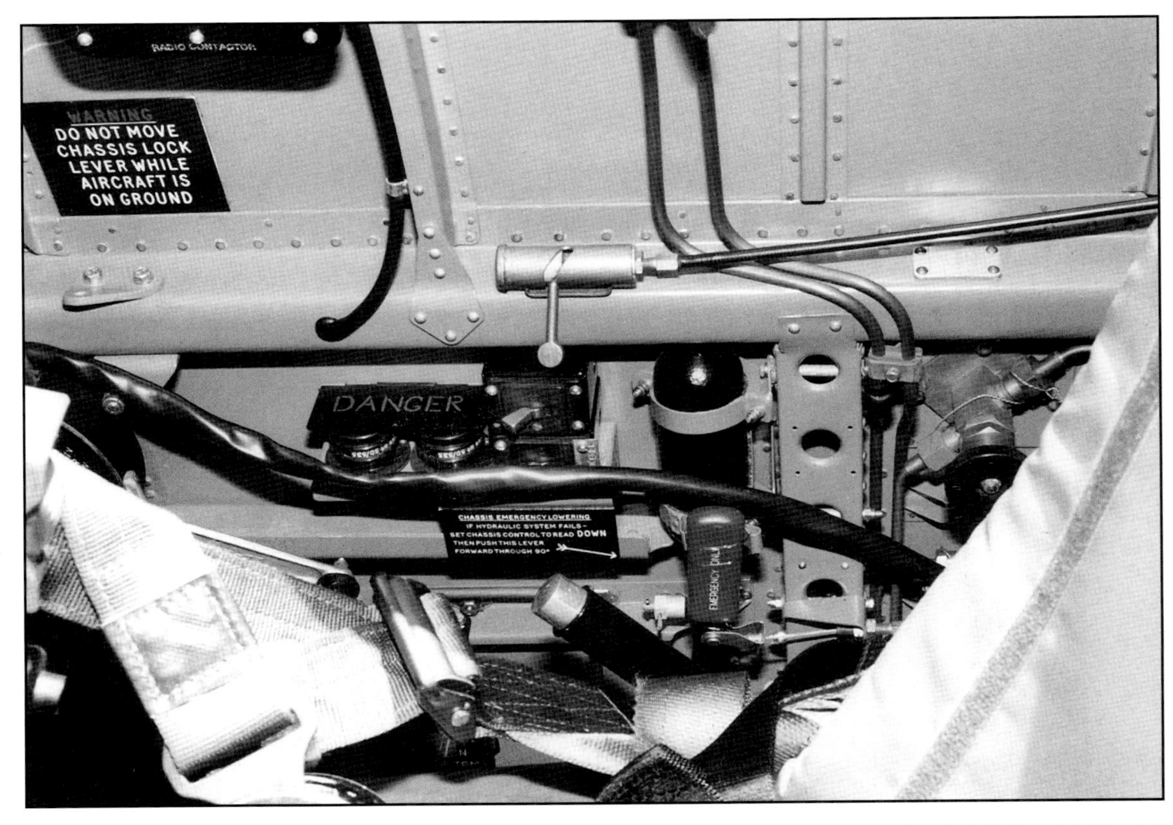

A further view of the starboard cockpit wall reveals the emergency oxygen system cylinder. (Chris Michell)

tually ending upside down. Although the damage was not beyond repair the decision to do so was negative as enough production aircraft were now available for trials purposes. The damaged airframe was not completely destroyed as major sections were retained by the photographic section at Farnborough for camera installation trials in both the wings and fuselage.

Although the prototype Spitfire came to an inglorious end it had flown some 200 hours and had achieved approximately 400 landings. The total cost of building the aircraft to contract was no more than £15,776, a very good deal for the taxpayer indeed.

The gauges and controls on starboard side of the pilot's pane were mainly concerned with engine and fuel system controls. (Chris Michell)

The technical specifications appertaining to the prototype Spitfire are as follows:

Wing Span 37 feet with a fuselage length of 29 feet 11 inches. Overall height over the tip of the propeller was 12 feet 8 inches whilst the wing area was calculated as 242 square feet. Flap operation was given as 57 degrees in the fully down position whilst the wings dihedral was set at six degrees with an aspect ratio of 5.67. Wing incidence at the root came out at +2 degrees which reduced to -1/2 degree at the tip.

The aircraft's Tare weight was found to be 4,082 lbs whilst the gross weight for the initial trials was calculated at 5,359 lbs although this included a fuel load of 75 gallons, seven gallons of oils, an allowance of 200 lbs for the pilot plus 436 lbs for service equipment and ballast in lieu of missing equipment. This was reduced slightly to 5,332 lbs during the 1936 flight trials. When the armament was fitted it consisted of eight .303 browning machine guns with 300 rounds of ammunition per gun.

Powerplants fitted to K5054 started with a Merlin Type C rated at 990 horsepower although this was later replaced by the Type F rated at 1,045 horsepower. Propellers flown on the prototype included a wooden two-bladed unit of 10 feet 6 inches in diameter as well as a Fairey three-bladed fixed pitch metal type plus a similar one from de Havilland, although this was of the two pitch variety.

Located on the starboard cockpit wall are the controls for locking the undercarriage and others for the electrical services. (Chris Michell)

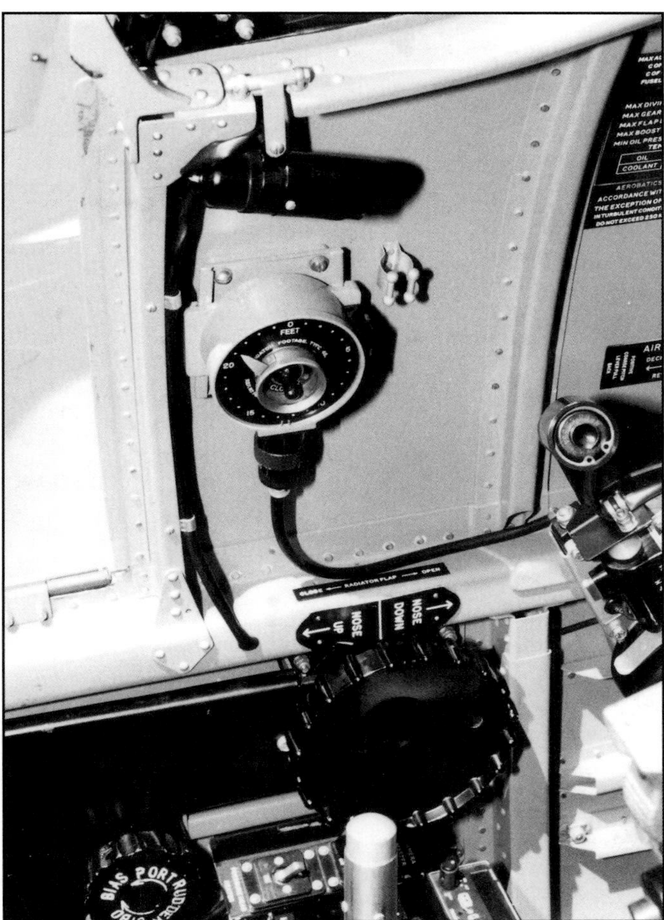

On the port cockpit wall are the trim controls for the rudder and elevators. (Chris Michell)

Close up of the undercarriage or chassis control lever located on the starboard side of the cockpit. (Chris Michell)

This view shows both the inner and outer faces of the main undercarriage fairings. (Chris Michell)

Seen from head on is the complete port undercarriage assembly plus leg fairing that was fitted to the greater majority of Merlin-powered Spitfires. (Chris Michell)

A sight not often seen is this immaculate view of the main undercarriage and its mounting plus the lock mechanism. (Chris Michell)

Although the undercarriage was hydraulically driven and came equipped with positive locking, such was the distrust of this new technology that pilots insisted that the early Spitfires had a visual method of undercarriage position indication as well as the cockpit lights. (Chris Michell)

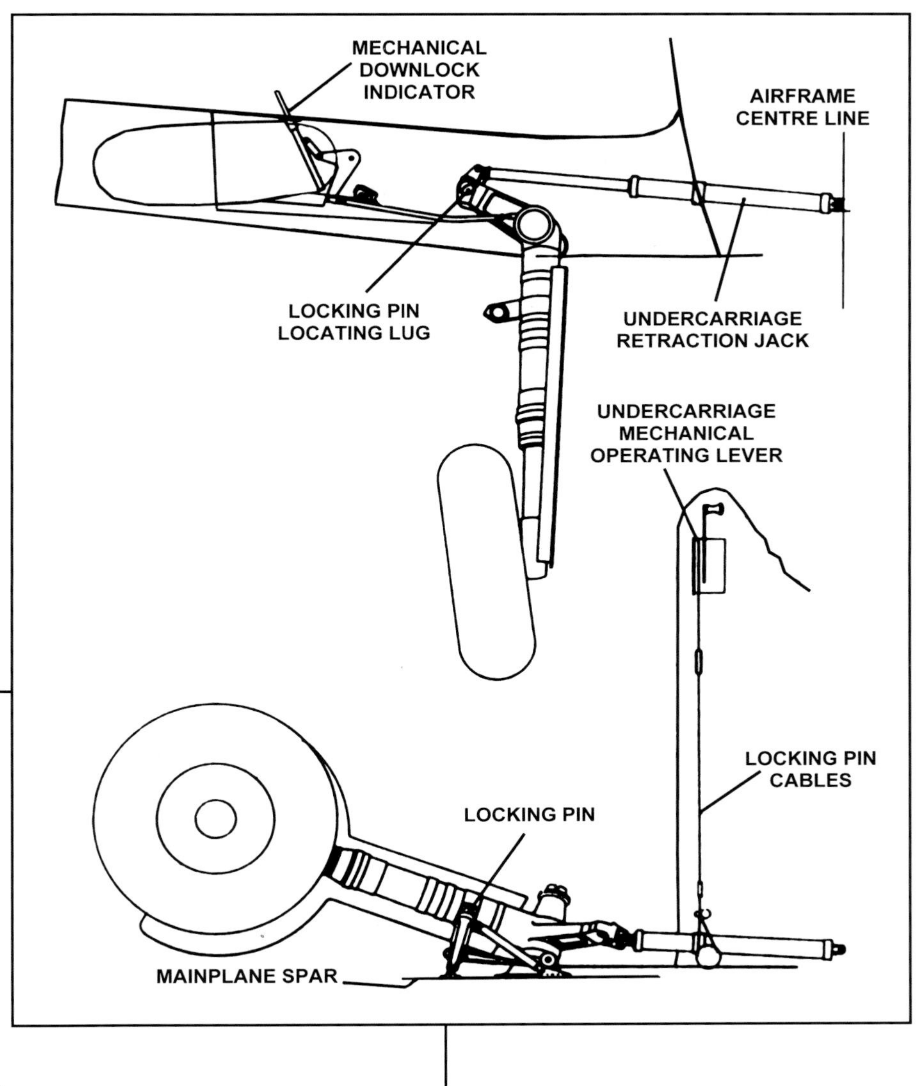

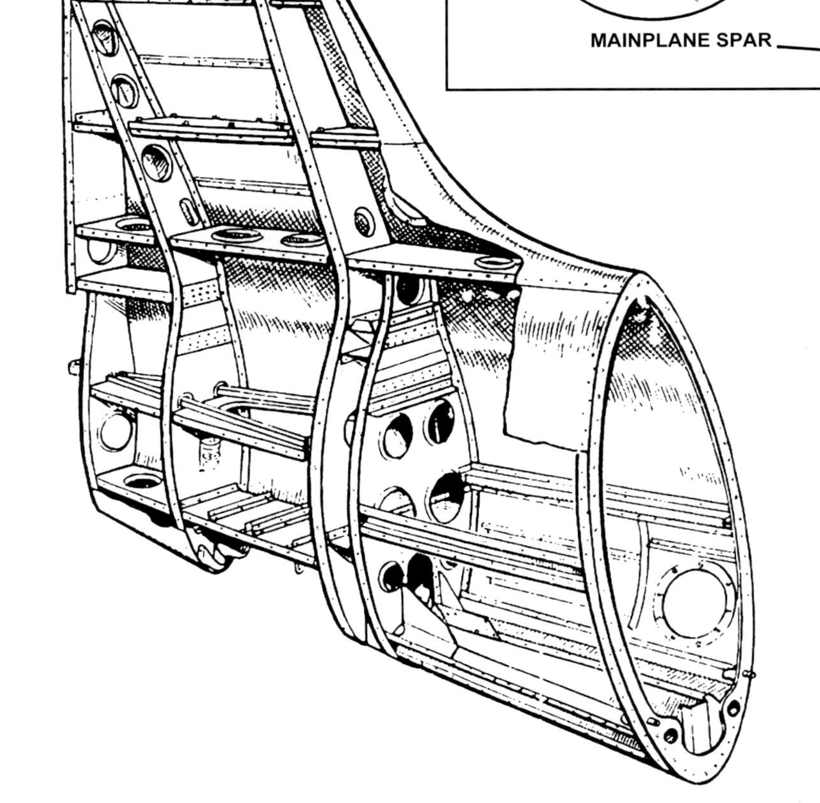

The aircraft's tail unit was quite frequently built by sub contractors such as Cunliffe Owen Ltd. Occasional accidents revealed that under certain circumstances the top mountings would fail, although the lower items seemed to retain their inherent strength. Modification action eventually reduced the problem. (Chris Michell)

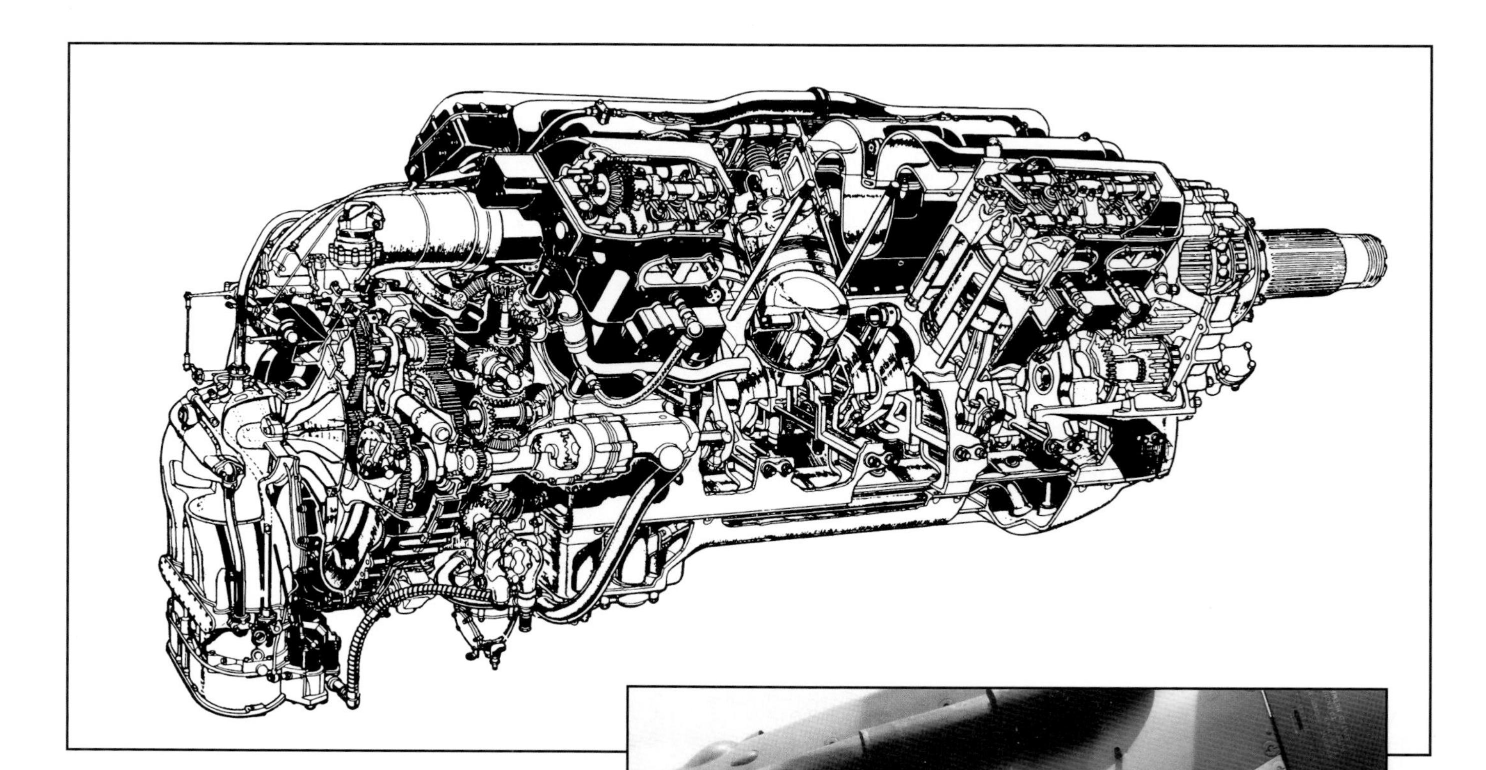

The heart of the Spitfire was the Merlin engine shown here in cutaway form. (Rolls-Royce Derby)

This view under the nose reveals the layout of the panels that surround the Merlin engine. Also clearly visible are the triple exhaust ejectors which were later replaced by blocks of six ejectors. (Chris Michell)

On the early Merlin-powered Spitfires the engine oil cooler was housed in a circular fairing under the port wing. (Chris Michell)

3 Into the Combat Zone

The Spitfire Mk.I and Mk.II

With the prototype safely in the air and achieving the majority of its projected performance targets the Air Ministry issued a revised Specification to cover the already ordered 310 production versions. Numbered Spec F 16/36 this document laid out in detail the changes that were required in the production Spitfire Mk.I. The greatest of these changes concerned the structure of the wing which was to be strengthened and stiffened to allow the aircraft to achieve a top speed of 450 mph indicated. Further changes saw the fuel capacity increasing to 84 gallons from the original 75 gallons. In order to improve landing stability, the flap travel was increased from 57 degrees to 85 degrees. Other areas that were revised were some parts of the production process which were changed from built up sections to castings and forgings. In common with most aircraft, the Spitfire prototype had literally been hand built therefore a full revision of all the blue prints was needed to allow for machine manufacture and in turn increase the rate of delivery.

Although Supermarine had promised in February 1936 that production could begin within fifteen months of the order being placed, in June it was obvious this was a very optimistic forecast and that October 1937 would be difficult to achieve. Much of the problem stemmed from the Supermarine production setup which encompassed no more than two small production facilities with a final assembly shed at Eastleigh aerodrome. Allied to this the company employed a very small workforce of approximately 500 persons and they were fully employed in building the Walrus amphibian whilst the parent company, Vickers, was busily engaged in manufacturing the Wellington bomber. Setting a precedence, the company subcontracted some of the work to another small firm, General Aircraft Ltd., which manufactured tail units.

This did help alleviate the pressure somewhat over the short term. However, a series of meetings throughout 1937 eventually decided that to increase production to an adequate rate, then four-fifths of

Contrary to the obvious explanation that this is a late-build Spitfire, HK856, this is in fact a Mk.I that had originally been sold to Turkey. It is pictured here at No. 1 M.E T.S based at El Ballah and wears the desert camouflage applicable to the zone. (Eric B Morgan Collection)

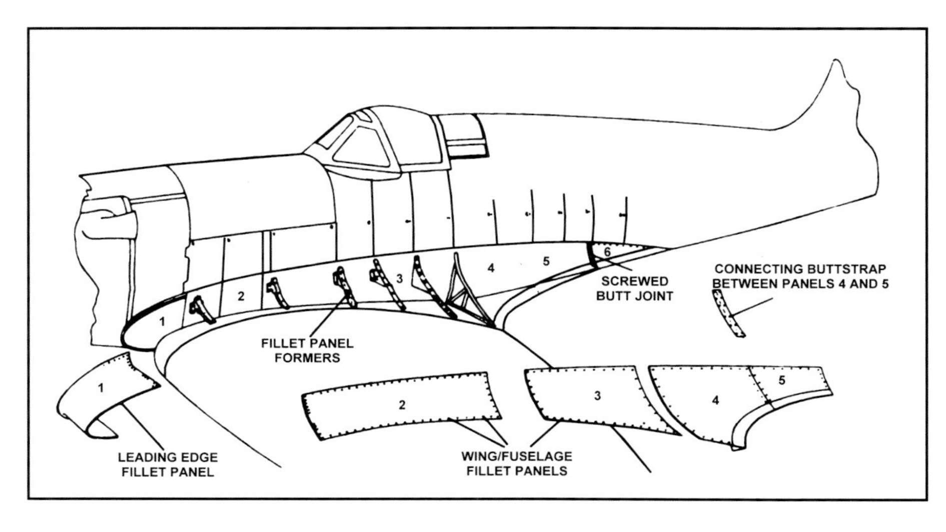

At the fuselage to wing interface Supermarine, in common with other aircraft builders, used shaped fairing panels to blend out the angular disparity between the two structures. As the location of the fairing supports varied from aircraft to aircraft they had to be fitted individually. (Chris Michell)

manufacture should be built away from Supermarines' Southampton base. This was finally agreed and all seemed settled until it was later revealed that many of the production drawings were not cleared for use and production difficulties arose again. Eventually the various problems were ironed out enough for the first production airframe K9787 to fly on 15 May 1938. The pilot was Jeffrey Quill who reported that the aircraft handled very well and was far more stable during landing due to the increase in flap range. Further testing at Farnborough revealed that the Spitfire had a possible top speed of 470 mph which was more than predicted. All was not completely

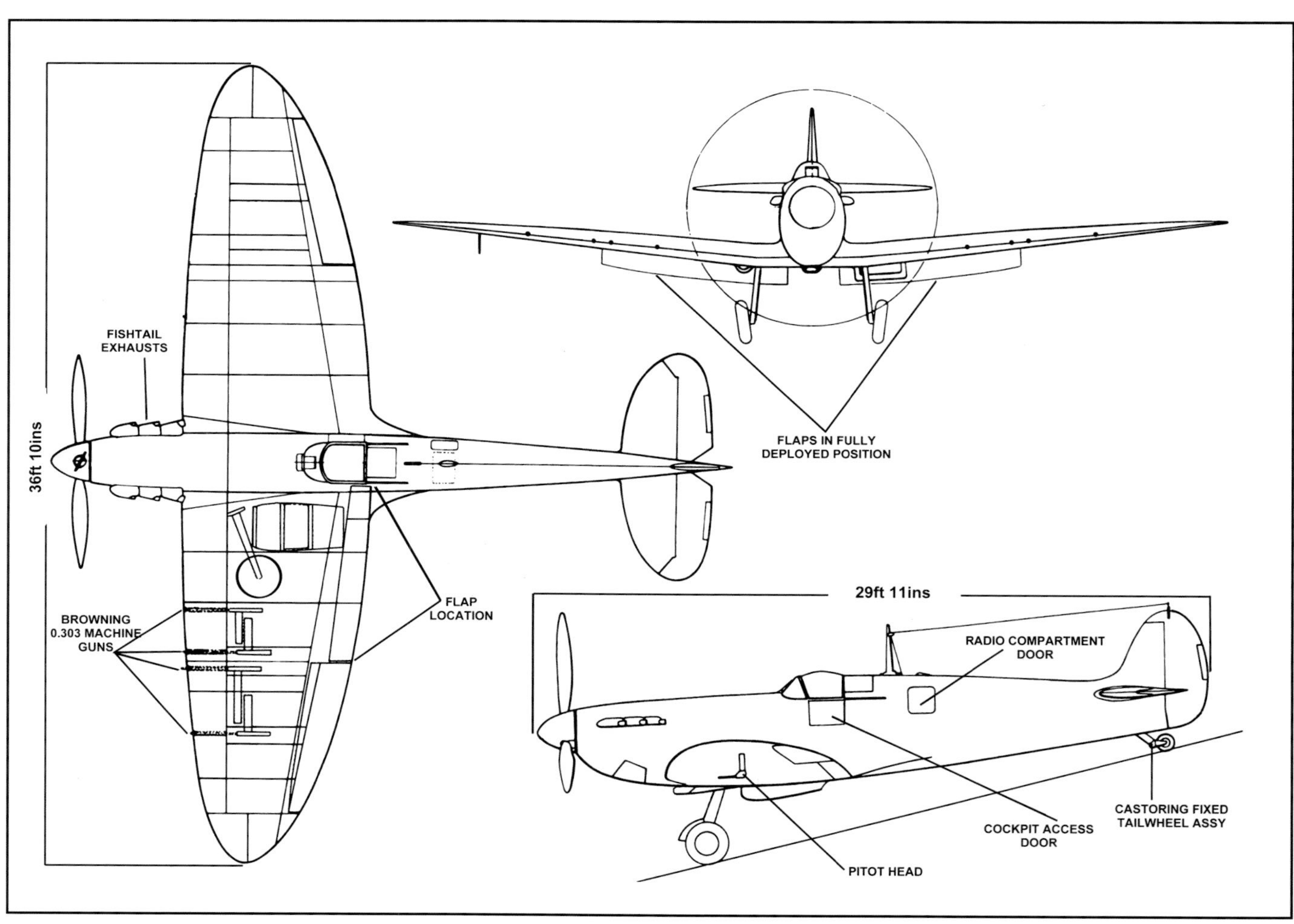

The general arrangement diagrams for the Spitfire Mk.I and Mk.II were similar in outline. The only variations that occurred were due to armament fittings. (Big Bird Aviation Collection)

well with the first Mk.I during flight testing, however, as it was discovered that in a dive above 400 mph the ailerons locked up. As the Spitfire had never been combat tested this was thought not to be a problem and would act as a protection against overstressing the airframe.

Deliveries of production Spitfires continued very slowly, the main problem area being the production of the wing assemblies which were far more complicated to build than the fuselage. At one point the production line had 78 fuselages, but only three sets of wings. The RAF received its first production Spitfire, K9789, on 4 August when No. 19 Sqdn based at Duxford began to reequip. Deliveries continued slowly with a total of 49 being in service by the beginning of 1939.

Whilst the problems of production were being ironed out, interest was being expressed by many foreign governments which had already seen the prototype perform. First off the mark in early 1937 was Japan which expressed a wish to purchase a single example for license production by Mitsubishi. Further expressions of interest also came from Belgium, Estonia, Holland, Lithuania, Switzerland, Turkey, and Yugoslavia. The first contract was signed with Estonia and covered the delivery of 12 aircraft, a similar number being ordered soon after by Greece. A further order later came from Portugal and was for 15 examples. Jumping the queue was France which requested a single example for consideration by the Armee d'el Air. Thus, in June 1939 the 208th airframe was flown to France complete with French markings. It was to be the only example to leave British shores before war was declared in September 1939.

Fortunately for the RAF production of the Spitfire had begun to increase so that at least ten squadrons were equipped and flying the type with another receiving its first examples. Altogether a total of 309 aircraft were in service with 71 in reserve. Whilst the Royal Air Force was equipping with the Spitfire, a variety of different trials were being carried out.

In order to extend the aircraft's range a selection of underwing tanks were tested, although they would only appear on the later reconnaissance versions. Another trial that would see service on production aircraft was that of the 20mm Hispano cannon, chosen because it had the best penetration of any weapon then available. One aircraft, L1007, was fitted with a pair of cannon and 60 rounds per gun. Initial testing was carried out at Martlesham Heath and later on the gunnery range at Orfordness. Although the cannon was seen as an advantage, its performance in these early stages let it down. During firing in high 'g' turns it exhibited a tendency to jam which occasionally meant that it shook itself apart. Development work would continue on the weapon to make it capable of withstanding the forces generated by a twisting wing instead of its original purpose of firing through the propeller boss.

Changes to the actual production aircraft had begun earlier in the run when the 78th Spitfire was delivered with a three-bladed de Havilland metal propeller in place of the earlier two-bladed wooden assembly. Further physical changes included a bulged canopy as standard and upperwing vents for the wing gun hot air venting system.

First combat operations for the Spitfire involved a scramble by No. 74 Sqdn which had been warned by local radar that a force of aircraft were proceeding up the Thames towards London. Anti-aircraft batteries confirmed that the aircraft

Spitfire Mk.I, X4257, is pictured at Boscombe Down undergoing testing. This side-on view shows the clean lines of the Merlin-powered Spitfire. (Eric B Morgan Collection)

were twin engined in configuration. In what became known as the "Battle of Barking Creek", the Spitfire squadron descended to engage with a force of single-engined fighters. Eventually the shambles was sorted out and revealed that not one aircraft from the Luftwaffe had been involved. Actual combat took place on 16 October when aircraft of Nos. 602 and 603 Sqdns engaged a force of nine Junkers Ju-88s attacking ships of the Royal Navy in the Firth of Forth. This time the RAF fighters shot down two confirmed and claimed one damaged.

Combat action for the Spitfires squadrons of the RAF was very infrequent during the remainder of 1939 and the early part of 1940. Unlike other aircraft types the Supermarine fighters were retained within the UK. In order that the pilots could gain some combat experience, the Spitfire squadrons began flying support missions over Belgium, Holland, and Northern France in support of the BEF and the French forces. It was in this theatre that the Spitfire first encountered its Luftwaffe equivalent, the Messerschmitt Me-109E, against which it acquitted itself admirably. The peak of these engagements came during the evacuation from Dunkirk after which there was a lull in hostile activities.

Whilst there was a lull in the fighting some further improvements were made to the in-service Spitfires. New aircraft were being delivered with armoured windscreens and variable pitch three-bladed propellers from either Dowty or Rotol. Already in-service aircraft also underwent these modifications as well as a fuel upgrade from 87 to 100 percent octane petrol. This gave no extra improvement in performance above 16,500 feet, but did improve performance below that height. Another much needed addition was IFF which would help ground radar differentiate between friend and foe. One further modification of note was also coming to fruition at this time, this being the cannon-armed Spitfire which was finally becoming effective. In-service usage was by No. 19 Sqdn which was operating six of the type. Once authorised for mass production the machine gun variant was redesignated the Mk.IA whilst the cannon version was designated the Mk.IB. By this time the Royal Air Force had 19 squadrons of Spitfires to call upon.

The Battle of Britain was to be the biggest air battle of the 1939-45 war. From the Luftwaffe point of view it was intended to destroy the air defence assets of the UK and to bomb the civilian population into submission. All was intended as a prelude to the invasion of Britain known as Operation Sea Lion. In opposition were the fighter forces of the RAF and some from the Fleet Air Arm all intent upon defending the British Isles.

The German forces initially launched attacks against their primary targets using Heinkel He-111 and Ju-88 bombers escorted by Me-109 and Me-110 heavy fighters. As each air raid was called up by radar detection both Hurricane and Spitfire squadrons rose up to repel the attackers. The Hurricanes went for the bombers due to their lower top speed whilst the Spitfires concentrated upon the fighters. Although both types managed many successes, the C-in-C Fighter Command Air Chief Marshall Dowding began to express doubts about the nation's survival. In fact, he is often quoted as saying, "Our young men will have to shoot down their young men at a rate of three to one." Some-

The first trials of overload tanks for the Spitfire involved fixed installations under the port wing. This is Mk.I, P9565, with a 40-gallon underwing tank. Although some aircraft used this extra in combat it was deemed as unacceptable in use as it reduced the aircraft's performance and manouverability. (Eric B Morgan Collection)

Seen from head on this Spitfire Mk.I is unique in that it has a pair of underwing overload tanks rated at 70 gallons each. The camouflage scheme under the wings consists of black and white to help differentiate between the aircraft of the RAF and the Luftwaffe. Although the installation was trialled by No. 602 Sqdn it was found wanting in service. (Eric B Morgan Collection)

how or other the fighter defence held up as did the various repair organisations upon whose efforts many of the frontline squadrons were reliant.

Possibly the month that saw the greatest fighting and the highest loss rate was August 1940. On one day over 110 Spitfire sorties were launched during which the type claimed 59 German aircraft damaged or destroyed for a loss of only five aircraft. In combat with the Me-109E, the Spitfire had some advantages, albeit minor, over its German rival. The Me-109E performance over a variety of altitudes was very similar, although the Spitfire did have a tighter and faster turning circle therefore the greatest advantage was mainly seen as tactical.

However, the German fighter did have one ability that the Spitfire did not. It could, when under attack, push its nose down into a bunt. The fuel system of the Daimler Benz engine was of the direct injection variety which allowed it to keep running under almost all circumstances. In contrast, the Merlin's fuel supply was governed by a float switch which would shut off the system should the Spitfire pilot try the same manoeuvre. The only way for the British aircraft to combat this was to roll over and dive after the other aircraft then perform an aileron flick to return to the right way up.

Another area that caused concern was the tendency of the ailerons to lock up at high speeds above 400 mph. Close investigation revealed that the fabric covering of the flying surface was ballooning out under pressure. To counteract this behaviour the ailerons of one aircraft were reskinned with a light alloy covering. This small change was to be of immense value to the fighter pilots of the RAF as it increased the Spitfire manoeuvrability at high speed by 100%. As the modification was such a success, an immediate crash programme was put in hand to implement this change as soon as possible across the whole fleet.

Combat experience being fed back to Fighter Command and Supermarine by pilots in the frontline squadrons was to result in some changes being made to the cockpit and windscreen assemblies. One of the greatest problems was the distorted vision caused by the curved side panels of the windscreen. It was recommended that flat side panels be installed which was later done. A greater area of concern was the fogging of the windscreens in a high-speed dive, although this was easily cured by the installation of a hot air louvre blowing on the screens. Escape by pilots from doomed aircraft was also giving rise to concern as many of those forced to bail out had reported struggling to clear the aircraft. Work was immediately put in hand by the design team at Supermarine to produce an easily jettisonable canopy which would appear on later variants.

The loss of fuel pressure in the inverted position was referred to Rolls-Royce, which responded by saying that the use of fuel injection had been considered but was rejected as it would have resulted in a loss of speed. Realising that solving this problem was of the utmost importance, development engineers at both Farnborough and Rolls-Royce began designing a new kind of carburetor that could function under negative G conditions although delivery would be sometime in the future.

Another area that was giving rise for concern was the performance, or lack of it, with the cannon-armed Spitfire Mk.IB. Initially the problems of jamming that afflicted the earlier aircraft were believed to have been cured. In service the opposite was found to be true as statistical figures drawn from the Battle of Britain reveal. On one day of the battle only the cannon on one of seven aircraft performed as advertised whilst later in that period a total of zero managed to perform. Further through the battle the success rate remained at approximately 25 percent. This obviously was inadequate as the commander of No. 19 Sqdn was quick to point out. In response, Fighter Command allowed the unit to swap its aircraft for Spitfires with machine guns that were in use with the Operational Training Unit at Hawarden.

Although the first venture of the cannon-armed Spitfire was not a success, further development work saw the revised Mk.IBs being issued to No. 92 Sqdn in November 1940. However, the fighting in this air battle had all but died down. Even though the Spitfire Mk.IB was now an operational aircraft, it had already been recognised that a greater spread of wing armament including machine guns would be needed to make the Spitfire an adequate mixed weapon fighter.

Spitfire Mk.I production ceased at Supermarine in March 1941 whilst the final batch was constructed by Westland Aircraft between July and September of that year. By the time manufacture had switched to later marques of the Spitfire, a total of 1,556 had been built. As for the aircraft themselves those that survived frontline usage went on to serve with various second line units such as No. 79 OTU in early 1941 where they were used in the conversion of tyro pilots to the type. Others found a second use as ground instructional airframes at various engineering training schools whilst others remained in the service of other second line units for use as station or unit "hacks".

The model of Spitfire that was to follow the initial variant was unsurprisingly designated the Mk.II. Externally there was very little difference between the two versions; the primary changes being confined mainly to the type of engine installed. The powerplant specified for the Mk.II was the improved Merlin XII which featured a Coffman cartridge starter plus a pressurised water glycol cooling system. Other changes were more of a cosmetic nature and covered alterations to the propeller spinner which was slightly blunter, plus the addition of a small blister shape

Very few Spitfire Mk.Is survive in preservation, this is one of them. Serialled AR213, this aircraft wears the codes No. 53 OTU although it is civilian-registered as G-AIST. In common with other early Spitfires in preservation this one has a late marque Merlin engine and four-bladed propeller. (Eric B Morgan Collection)

on the starboard side of the engine cowling which covered part of the cartridge starter gear train.

Production was centered upon the new facility at Castle Bromwich near Birmingham. Construction had begun in mid 1938 of the purpose-built factory which was to be managed by the car manufacturer Morris Motors. Although the production line was equipped with the best machinery available and was intended for operation by semi-skilled labour, the intention to build Spitfires like cars was to cause extensive problems as the planning involved had not taken into account the possible alterations that could be required by the end user, the Royal Air Force. Part of the problem being experienced by the facility was the numerous amount of modifications that were called for as experience of air combat with the Spitfire increased. This resulted in the facility producing components but no complete aircraft.

With tensions increasing due to the build up of German forces across the Channel, pressure was brought onto the Morris Motors management to actually start building complete aircraft prior to the battle ahead. By May 1940 it was obvious that Germany was about to launch its offensive. To get the production lines moving Vickers assumed control of Castle Bromwich. Initially ten airframes were completed by using major components built at the Supermarine factory. With the first hurdle of production completed, the amount of aircraft manufactured began to increase. Vickers used a hybrid method of manufacture which required that the basics were carried out by the semi-skilled workforce whilst the skilled personnel at both Castle Bromwich and Supermarine embodied the necessary modifications.

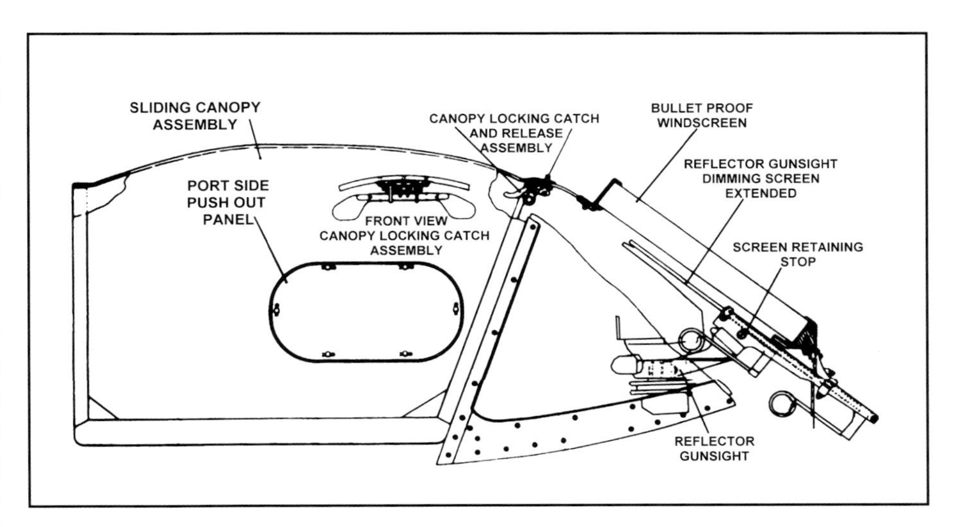

The hood fitted to the earliest Spitfires was flat topped, however combat experience dictated that it became of the blown variety as shown here. (via Owen Morris)

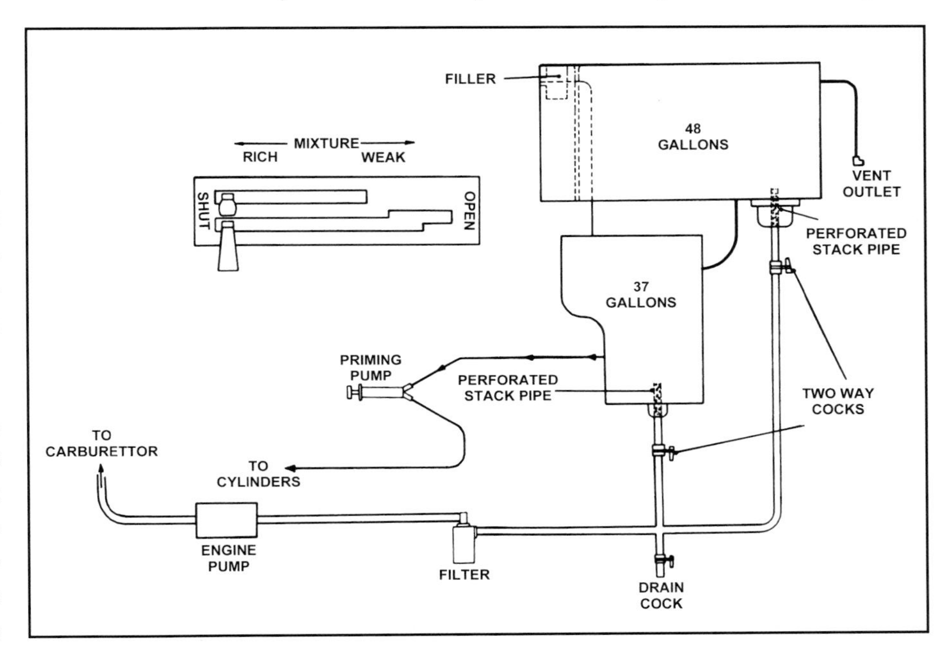

The design and layout of the fuel system in the early Spitfire marques was fairly simple in nature. As extra internal and external tanks were added so fuel system management became more critical. (Big Bird Aviation Collection)

First deliveries of the Spitfire Mk.II began in August 1940 when No. 611 Sqdn based at Digby, Lincolnshire, began to equip with the type. In the next month Nos. 19, 74, and 266 Sqdns also received the Mk.II and all four units took part in the latter stages of the Battle of Britain. In a similar circumstance to the Mk.I Spitfire, the latest version was delivered in two versions. The Mk.IIA was the standard aircraft armed with eight Browning machine guns, whilst the Mk.IIB was equipped with a pair of Hispano cannon plus four machine guns. Altogether a total of 921 Spit-

Spitfire Mk.I, L1090, was photographed just prior to the outbreak of war in 1939. It is unusual in that it has 15-inch, two-colour roundels on the fuselage. (Eric B Morgan Collection)

fire Mk.II aircraft were built at Castle Bromwich which included 751 of the eight gun version whilst the remainder were of the cannon armed Mk.IIB. Production of this version continued until March 1941 when the improved Spitfire Mk.V began to be built.

During the service of the Mk.II some were fitted with a fixed tank under the port wing containing an extra 40 gallons of fuel in order to increase the type's range. Examples were used by Nos. 66, 118, and 152 Sqdns, but were found to be lacking in speed and climbing performance and thus were not popular with squadron pilots. A further more productive use was found for retired Mk.II's when they left frontline service in 1943. From the 900-plus built, a total of 50 were fitted with the more powerful Merlin XX and Type E air sea rescue survival packs. This version was designated the Spitfire Mk.IIC and was used for air sea rescue missions over the various waters that surround the United Kingdom. Operationally this variant was flown by six squadrons stationed at strategic points around Britain. One other modification that was trailed on a Spitfire Mk.II was the improved Mk.II deflector gunsight which had improved gyros fitted and allowed more accurate offset gunnery.

In a similar manner to the earlier Mk.Is, the later versions moved onto secondary duties from 1944 when they were moved to the various OTUs to train new pilots. Those not employed upon these duties were found a new role as ground instructional airframes.

In appearance the Spitfire Mk.II was the same as the earlier Mk.I. Clearly visible are the fish tail exhausts, three-blade propeller assembly, oil cooler, and radiator fairing. (C P Russell Smith Collection)

SPITFIRE 4 MK.V

CLIPPED, CLAPPED, AND CROPPED

The next version of the Spitfire to be built was originally seen as a stopgap between the Mk.II and the planned Mk.III, which was seen as the definitive version of the Merlin genre. In its stead and with the cancellation of the Mk.III came the Mk.V that was to be built in large quantities and in four different versions. The requirement was placed in the summer of 1941 and was in response to the high-altitude flights being carried out over Britain by Junkers Ju-86P reconnaissance aircraft of the Luftwaffe which were operating at altitudes in excess of 38,000 feet virtually untouched.

In order to reach these raiders, a more powerful version of the Spitfire would be required. This was to be the Spitfire Mk.VI which featured a pressure cabin and a boosted engine. However, delays in development meant that a more versatile aircraft and one that could be produced quicker was needed in a hurry. This in turn would need an improved engine which appeared in the form of the Merlin 45. This was bolted onto a standard Spitfire fuselage whose Frame 5 mounting bulkhead was strengthened to accept the new powerplant.

The first production examples were converted from Spitfire Mk.Is and IIs already on the production line. First flight of a Spitfire Mk.V, X4922, was undertaken by test pilot Jeffrey Quill on 20 February 1941. The second aircraft, K9788, was used as the Merlin 45 testbed at Boscombe Down although it was flown without most of its operational equipment. Even so the Spitfire had gained another 14 mph in speed plus a measurable increase in performance in the areas of climb and high-altitude speed. Both of these aircraft were designated as the Mk.VA and featured what became known as the "A" Type wing which had the installation mountings for eight Browning machine guns. Only 94 of this version were built before production switched to the more capable Mk.VB. Unlike the Mk.IB, this new version of the Spitfire was not only armed with a pair of 20mm cannons but also featured four machine guns.

Production of the Mk.VB was divided between three facilities at Westlands, Supermarine, and Castle Bromwich and started in earnest during 1941. Deliveries to the RAF began in February 1941 when No. 92

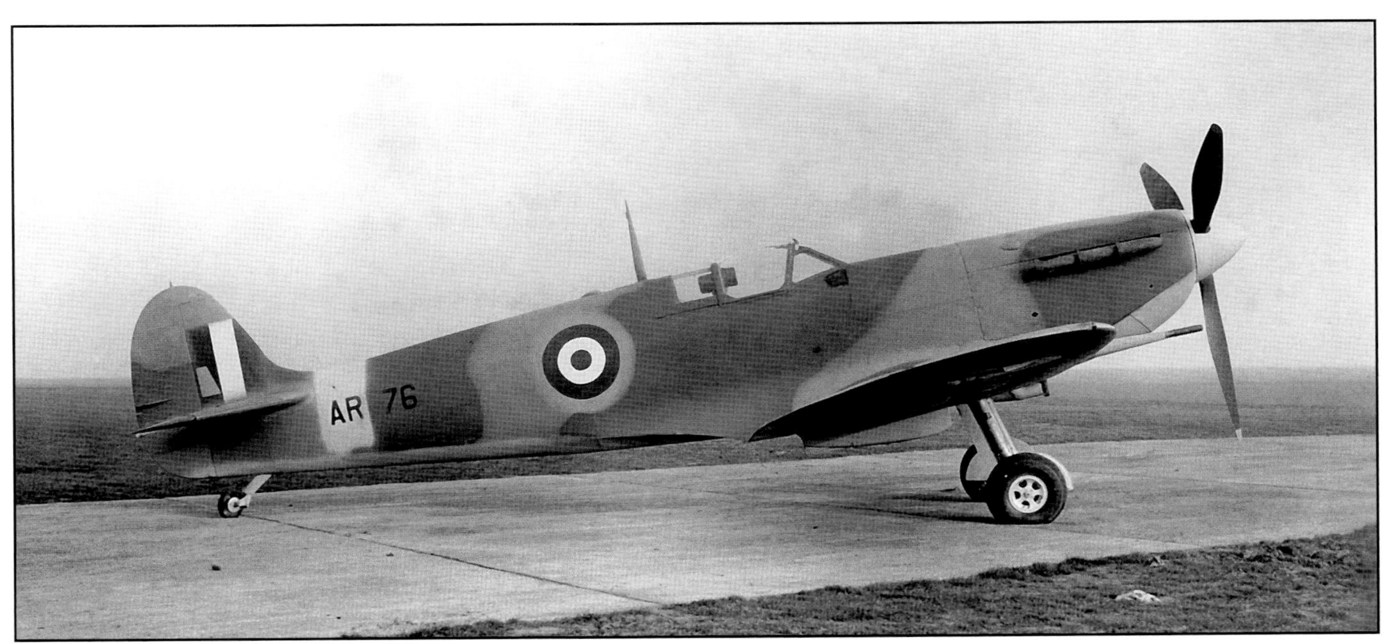

Photographed in December 1941 Spitfire Mk.VB was typical of the marque. AR376 initially served with No. 303 (Polish) Sqdn and survived the war. It was withdrawn from use in January 1945. (C P Russell Smith Collection)

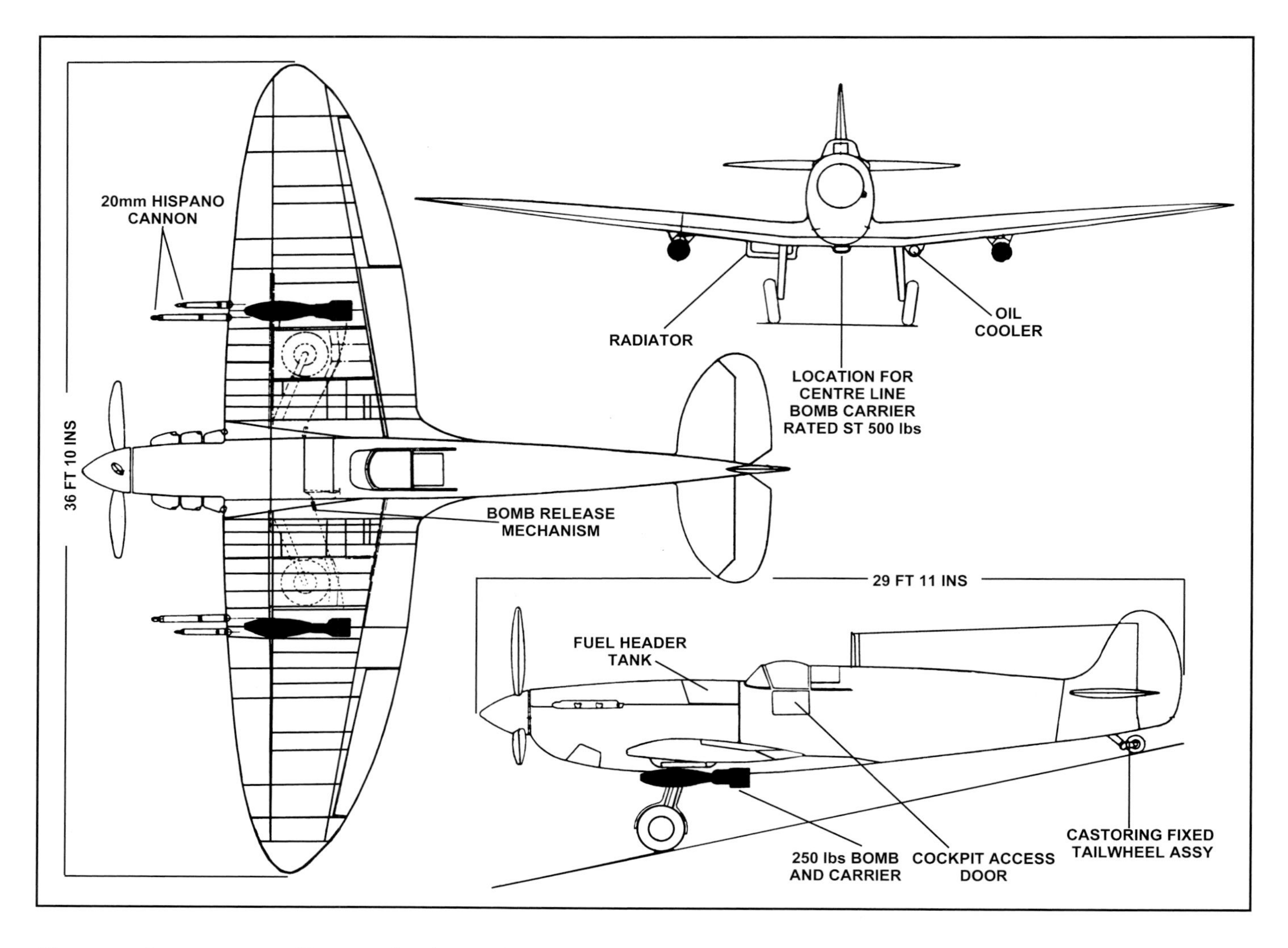

The general arrangement diagrams for the Spitfire Mk.V were many and varied. Shown here is that for the LF.VB complete with wing bomb load (see below for details). Other diagrams covered the clipped versions plus those modified for other duties. (Big Bird Aviation Collection)

Sqdn based at Biggin Hill received its first examples. Other units followed quickly with No. 91 Sqdn equipping in March, Nos. 54 and 603 in April, plus Nos. 74, 111, 609, and 611 during May. Continued high-volume manufacture meant that virtually all of the fighter units within the sphere of Fighter Command were equipped with the Spitfire Mk.V.

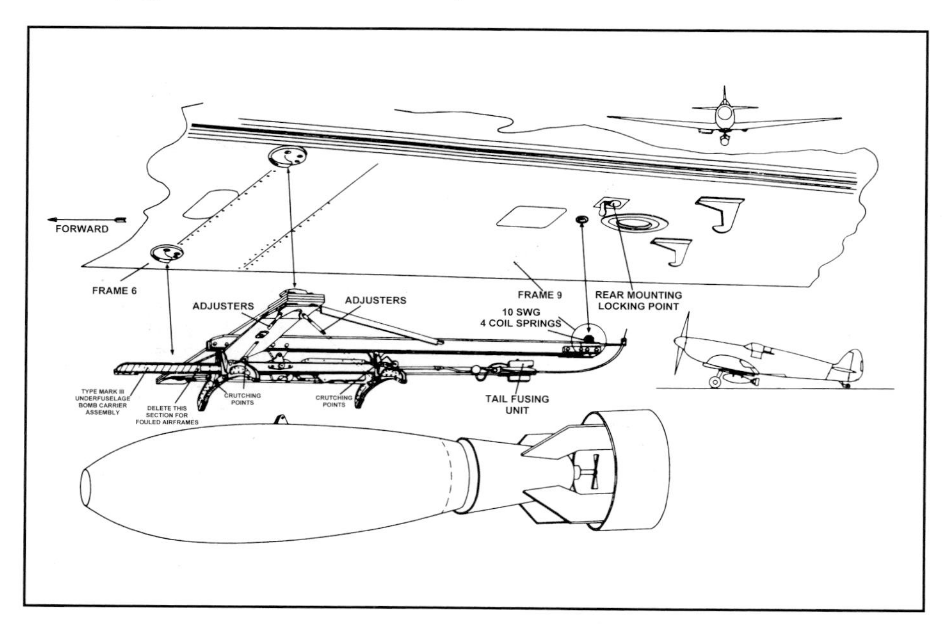

To increase the Spitfire's attack potential a centre line bomb carrier was developed. After trials it was cleared for the carriage of 500-lbs and 250-lbs GP bombs. (Big Bird Aviation Collection)

As these units were training on the new version, there was a change in strategic policy. Fighter Command would be carrying out offensive operations instead of those of a more defensive nature. Many of these missions involved flying fighter wing missions over France and Belgium as well as bomber escort sorties to targets within those same countries. As this involved longer mission times over a greater range, the lack of a safe fuel margin meant that the development of jettisonable external fuel tanks was deemed a high priority. Eventually development and trials work culminated in the production of 30- and 45-gallon underfuselage slipper tanks.

The next version of the Spitfire to appear was the Mk.VC which, although similar in outline to the Mk.VB, featured an entirely new wing which became known as the Universal type wing. All the changes were internal and featured greater strengthening of the structure plus the ability to tote either a pure machine gun armament or the more preferred two cannon and four machine gun arrangement. The cleverness of the redesign allowed an increase in the amount of ammunition available for the cannon from the original 60 rounds to 120. Another minor change was the slight raking forward of the main undercarriage in an effort to improve stability on the ground.

With the fighter units of Fighter Command now fully equipped with the Spitfire Mk.V it was deemed time for other units overseas to receive the new aircraft. The first theatre to gain the Mk.V was the Mediterranean Theatre of Operations for missions over Egypt, Libya, and Malta. Prior to despatching any aircraft overseas, trials were undertaken on a UK-based aircraft, X4922, in an effort to filter out the major problem in the region — dust. The answer was seen as a large dust filter fairing under the aircraft's nose. Although efficient, the size of the fairing caused a reduction in ram air pressure to the carburetor pressure which in turn reduced the aircraft's top speed by about 8 mph.

To convert a Spitfire for tropical use required at least 26 modifications to be made, most of these entailed installing much modified lower engine access panels to allow the installation of the filter and its associated pipework. This new version was given the company designation Type 352/6.

Another modification that made its appearance at this time was an enlarged underfuselage ventral tank which contained 90 gallons and was intended for ferry use only.

The deployment of the tropicalised Spitfire began on 7 March 1942 when Operation Spotter began. This

Although preserved as a Spitfire Mk.V this aircraft, part of the Battle of Britain flight, has been fitted with a late marque Merlin and four-bladed propeller. AB910 wears the codes of its first operator, No. 92 Sqdn. (Ray Deacon)

The extent of the amount of flap deployment available to the Spitfire is exemplified by this underneath shot. (Dave Stewart)

involved loading 15 aircraft aboard the aircraft carrier, HMS *Eagle*, from whose deck they would depart opposite the coast of Algeria. The destination was the Island of Malta some 660 miles away. The first attempt at launching was aborted due to fuel system and engine problems which required the attentions of a Supermarine specialist to overcome.

Once the Spitfires were launched, the carrier returned to Gibraltar to collect another batch of Spitfires for onwards transition to Malta. Although this method of delivery was successful the total of aircraft delivered was not really enough to make a great difference to the defence of the Island. So that a significant amount of aircraft could be delivered the British government prevailed upon their counterparts in the United States to lend support in the shape of the carrier, USS *Wasp*.

Ground crew service the Merlin powerplant of Spitfire Mk.IX EN455 'somewhere in Italy'. Note the battery trolley in front ,better known as a trolley acc, which is used to start the aircraft in preference to its own. (Big Bird Aviation Collection)

Airborne on a combat mission from Malta is Spitfire Mk.VC(T), JK925, fully modified for Middle East service. Given the weather conditions it is not surprising that the pilot is flying with the canopy open. (Big Bird Aviation Collection)

Agreement was forthcoming therefore Operation Calendar began on 20 April when both carriers launched a total of 47 Spitfire Mk.VCs. The next mission, named Operation Bowery, saw a further 64 Spitfires delivered using this method. To ease the loading of each Spitfire the wingtips were removed, although they were refitted once disconnected from the crane. All were armed with four cannon and came complete with camouflage suited for operations over dry and dusty landscapes. With such a significant increase in the islands air defence the tide of air attacks was turned and a measure of air superiority was established. Further deliveries of small quantities of aircraft helped to reinforce the garrison and replace any Spitfires that had been lost in combat.

Once the Spitfire Mk.V had become established in service it underwent a series of modifications. One of the first brought to the fore was the incorporation of a carburetor that was capable of continuous operation whatever G forces were applied to the airframe. The new carburetor was first installed on the uprated Merlin 50 that began to appear in the Mk.V in late 1941. A further modification to improve performance was a change from the original de Havilland unit to that from Rotol. This was slightly heavier than the original assembly and required that a counteracting ballast weight be fitted in the rear fuselage. However, it did show a marked increase in performance above 20,000 feet and raised the operational service ceiling to 39,600 feet. Deliveries were delayed of this unit as they were required for installation in the high-altitude variants.

One already approved modification that was failing to impact upon the production lines was that of metal covered ailerons. Most airframes were still being fitted with the earlier fabric-covered items with a subsequent loss in performance as a consequence. When some Spitfires were handed over to the USAAF, the first thing that was done to any of the unmodified aircraft was to cover the ailerons in thin sheet plywood. This was not as effective as alloy skinning, but was found to be better than fabric.

Although Fighter Command clearly regarded the Spitfire in all its forms as an air defence fighter, the hierarchy was astute enough to realise that an extension of the types weapons carrying ability would be an advantage. To that end extensive trials were carried out with a centreline bomb crutching unit that eventually cleared the standard 250 lbs and 500 lbs for service. During the same period, trials were also carried out with the flaps modified to act as dive brakes being operated by compressed air. The trials were deemed a success although the system was not adopted for service use.

This Spitfire Mk.VIII, JF814, is marked with the personal initials of Air Vice Marshal W F Dickson. As this is a late build airframe it has a pointed fin and retractable tail wheel. (Big Bird Aviation Collection)

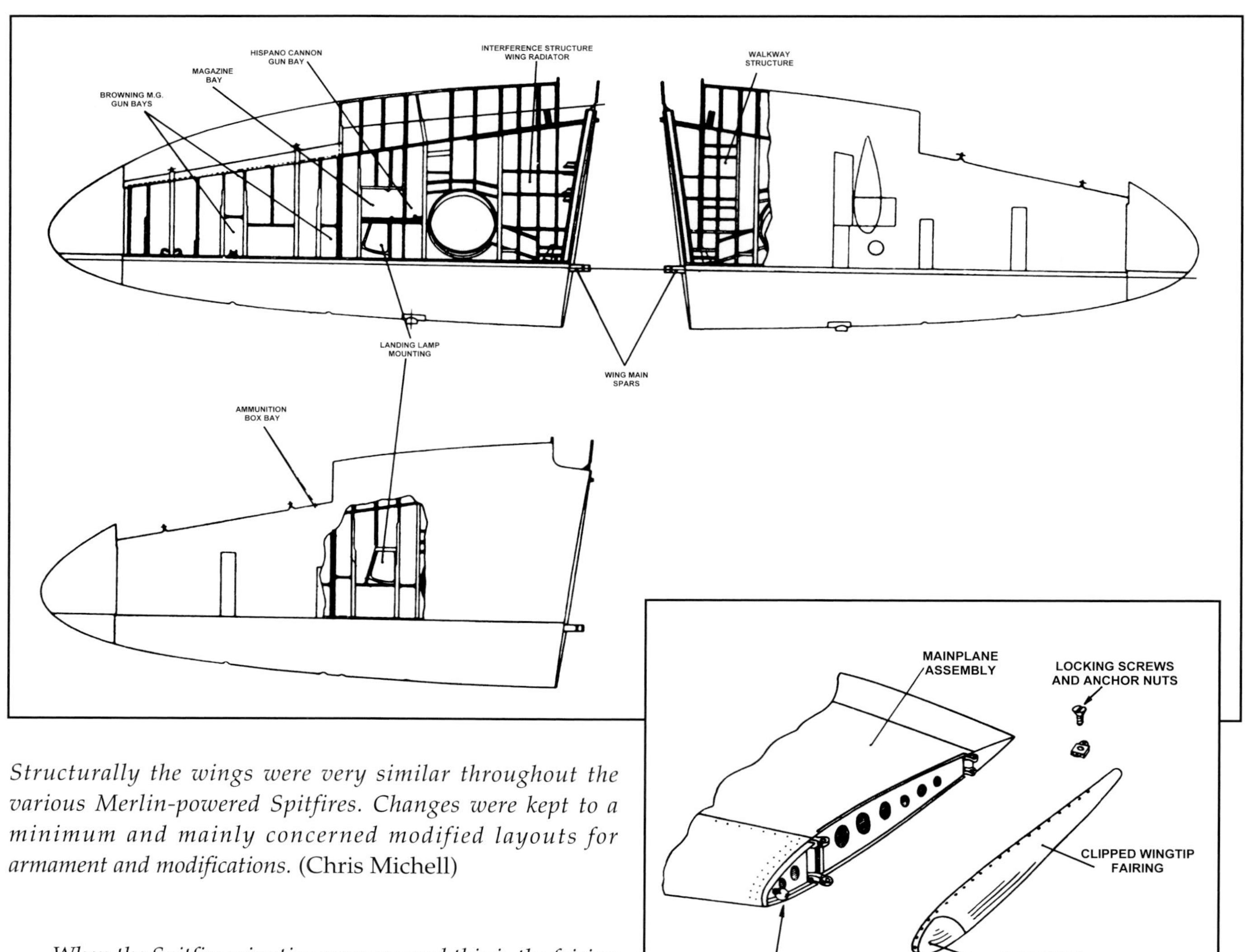

Structurally the wings were very similar throughout the various Merlin-powered Spitfires. Changes were kept to a minimum and mainly concerned modified layouts for armament and modifications. (Chris Michell)

When the Spitfire wingtips were cropped this is the fairing that was installed in the place of the standard assembly. (via Owen Morris)

The ability to carry a bomb load was grasped by commanders in the field as a way to extend the Spitfire's capabilities. Thus, it was not unusual to find fighter sweeps where 50 percent of the assigned aircraft were carrying a bomb load. An extension of this modification was undertaken in the MTO where mounts were installed under the wings. This re-establishment of an original requirement from the specification of 1934 allowed for the carriage of two 250-lbs bombs and turned the Spitfire into an effective ground-attack aircraft.

Whilst the performance and capabilities of the type were being improved, an important safety requirement was also being addressed. During combat it had been discovered that the lower fuselage fuel tank was very vulnerable to enemy weapons fire. Pilots had reported that when punctured it burnt very quickly and meant that pilots had to depart the damaged aircraft quickly or risk suffering severe burns. Originally a form of self sealing, either internal or external, was studied as a possible cure. Both were rejected as unworkable as the internal would reduce the fuel load to unacceptable levels whilst the external could not be fitted due to the lack of clearance. The answer was to strengthen and seal the front bulkhead which would improve the chances of pilots trying to abandon their damaged aircraft.

Further modifications were forced upon the designers of the Spitfire due to the appearance of the Focke-Wulf FW-190. This new addition to the frontline strength of the Luftwaffe was found to be faster, a better climber and diver, and had an improved rate of roll which put the Spitfire Mk.VC at a disadvantage. Fortunately for the RAF they had managed to capture an intact example when one landed by mistake in Kent. After a full evaluation by the A&AEE and the RAE the results were passed onto both Rolls-Royce and Supermarine.

The first solution that was proposed was to install the underdeveloped Merlin 61 which had a two-stage supercharger. However, this was some months away from full production. In the meantime some modifications were undertaken on the airframe to improve performance. The most noticeable of these was to crop the wingtips, which involved removing the original pointed assemblies and replacing them with blunter fairings. Under testing by the AFDU it was found that the clipped Mk.V had a slightly improved top speed, but had a marked increase in its rate of roll. A further change was the fitment of the Merlin 50M engine which was an upgrade of the original Merlin 45 with a cropped supercharger impeller. This modification allowed a maximum of +18 lbs boost at an altitude of 5,900 feet and gave the Mk.V a top speed of 350 mph which was almost a match for the FW-190. The official designation for this version was given as the LF.V, although it was more commonly known as the "clipped, clapped, and cropped" Spit. The reason for the "clapped" arose due to the age of the airframes that these modifications were applied to.

One theatre of operations that used the LF Mk.V extensively was that based in Malta. Originally the aircraft were intended purely for air defence. However, as the tide of war had begun to swing in the allies' favour the Spitfires based on Malta found their area of operations changing.

In August 1942, No. 126 Sqdn began flying combat operations over Sicily, its Spitfires having been modified to carry a pair of 250-lbs bombs under the wings. During each mission the pilots were briefed to attack targets of opportunity and to maintain a positive air presence over the island.

Changes were also afoot to improve the method of delivery as the use of aircraft carriers was expensive. The alternative was to install a modification kit which consisted of an enlarged oil tank below the nose, a 29-gallon fuel tank behind the pilot's seat, and a 170-gallon jettisonable fuel tank below the fuselage. This gave the Spitfire a total of 184 gallons of fuel which was enough for the aircraft to fly from Gibraltar to Malta, a distance of some 1100 miles.

Tied down for engine runs using enlarged chocks and a tail tie down strap this is a Spitfire Mk.VC. This aircraft still retains the standard wing tips which were later clipped for in-service use. (C P Russell Smith Collection)

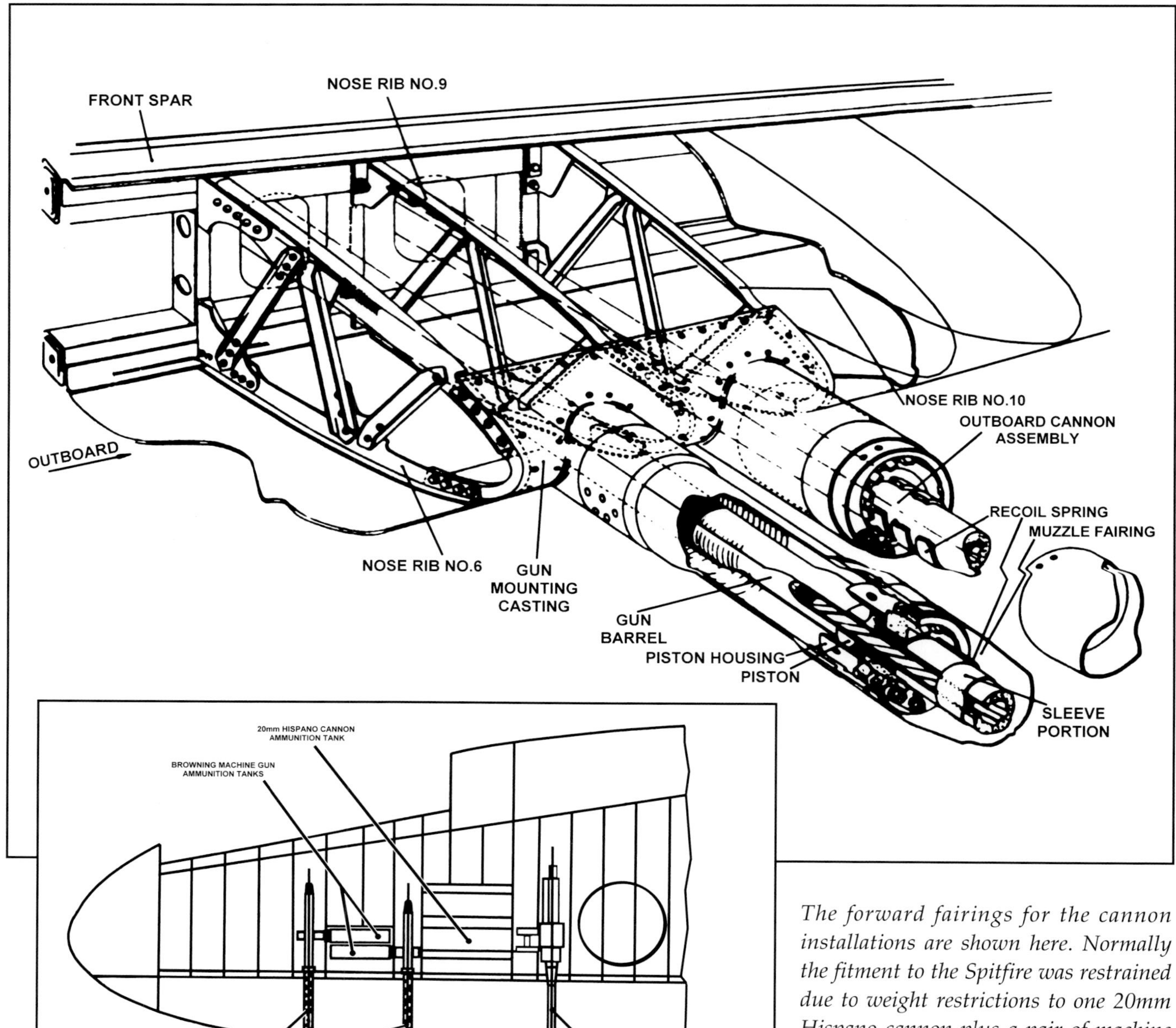

The forward fairings for the cannon installations are shown here. Normally the fitment to the Spitfire was restrained due to weight restrictions to one 20mm Hispano cannon plus a pair of machine guns per wing. (via Owen Morris)

This close-up arrangement diagram is applicable to the Spitfire Mk.VB and the later Mk.VC. (Chris Michell)

In order to reduce the extraneous weight of the Spitfire the armament was reduced to a pair of machine guns. Upon arrival in Malta each aircraft was stripped of its modification kit and the required armament and tropical filter was fitted. This particular assembly was modified by engineers at No. 103 MU, Aboukir, who managed to reduce the size of the original after diligent efforts. Although it was smaller it was found to be just as effective. It was to be installed in a slightly modified form from the outset on production Spitfire Mk.IXs.

Shortly thereafter the Spitfire was designated the Mk.VC(T) and these were delivered to the Middle East forces in Egypt to equip Nos. 92 and 145 Sqdns. A total of 50 airframes were despatched by convoy in December 1941 in knock down kit form. Upon arrival this meant that there would be a delay in aircraft

reaching the squadrons as they would need to be reassembled and thoroughly tested before delivery. Realising that the attrition replacement airframes, nominally set at 10 per month, could not be delayed due to the need to fully reassemble them, instead they were partially stripped and carried as deck cargo. Although not ideal it did speed up delivery.

One other major modification involved just a single airframe, BP985. This was reworked at depot level in May 1942 for one specific task — shooting down the quartet of Junkers Ju-86P high-altitude reconnaissance aircraft that were monitoring fleet movements in the Egypt area. Although the Spitfire was capable of nearing the 37,000 feet altitude, it was unstable as a gun platform at this altitude. The rebuilt Spitfire had a modified Merlin 46 engine installed. This had an increased compression ratio as it had been impossible to boost the supercharger gear ratio. To assist in increasing compression the engine block was rebored and modified liners installed. Instead of the normal three-blade propeller, a four-blade de Havilland Hydromatic unit was installed. Also fitted in place of the original equipment was an undernose Aboukir filter incorporating an oil tank.

Once the performance had been enhanced, further work was undertaken on the airframe which mainly involved the removal of any excess weight possible and the reduction of the installed armament to just two Browning machine guns. To further improve high-altitude performance a pair of extended wingtips built by the depot were fitted. The modified Spitfire was deemed successful as its presence chased the Luftwaffe spies away.

Whilst the Spitfire was scoring successes it was also suffering accidents, some of which needed extensive investigation to uncover the cause. One of the most confusing involved the ailerons, malfunctions of which were causing aircraft to crash. To complicate matters, in some crashes a wing would detach whilst in others the tail unit would become detached at frame 19 in the rear fuselage. Close inspection of in-service aircraft revealed that the lower tail unit mount was strong enough whilst the upper mounts would need increased strengthening. Many theories were put forward to sort out the aileron problem including increased cable tensions, counter balance weights, and a form of inertial damping. The eventual solution was to put a small amount of negative aileron droop onto each surface plus increasing the monitoring of the rigging of the aileron circuits.

Further problems also arose with distortion of the upper mainplane skin above the wheel wells. The answer was to fit external strengthening stringers, the work being carried out at a local level for minor distortions whilst transfer to a depot was required for more major repairs. Overstressing for various reasons was also causing some concern. In some cases manoeuvring caused failure of the starboard tailplane whilst firing of the cannon was causing problems with the oxygen system, which either failed completely or became uncontrollable. In either case it caused problems for the pilot who either became incapacitated or was forced to reduce altitude rapidly. Another series of accidents that perplexed the investigators was a sequence of accidents where the wings were coming off the aircraft during combat manoeuvres. Inspection of the crashed airframes suggested that there was a failure in construction which was tempered by improving inspection during manufacture.

On 6 August 1942 Spitfire Mk.VA W3719 of No. 66 Sqdn suffered an engine failure which resulted in a forced landing in occupied Europe. This airframe was originally built as a Mk.I before conversion to the later marque. (C P Russell Smith Collection)

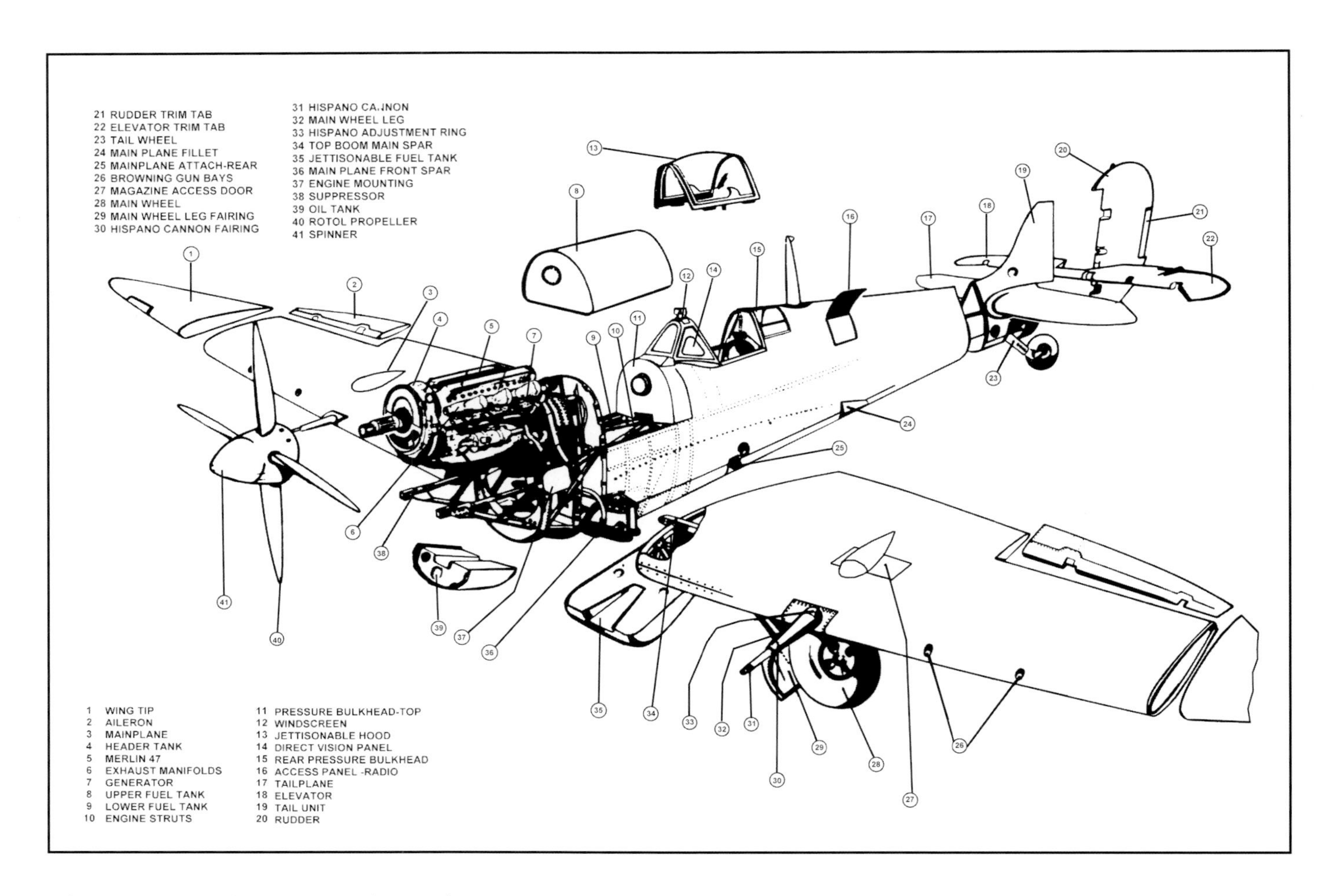

The major parts that comprise the Spitfire are shown here. Although this is dedicated to the high-altitude interceptors most Merlin-powered aircraft were similar. (Big Bird Aviation Collection)

Spitfire X4922 was originally constructed as a Mk.I. By 1 January 1942 it had been converted to act as the prototype Mk.VA. Trials with the Vokes tropical filter followed, these being complimented later by evaluation of the 90-gallon overload tank. X4922 is seen here complete with filter assembly and external fuel tank. (C P Russell Smith Collection)

Spitfire W3373 was built at Castle Bromwich as a Mk.VB. It is seen here awaiting delivery to its first operator No. 609 Sqdn in June 1941. Visible are the rear view mirror above the canopy frame and the emergency access panel for the pilot in the canopy hood. (C P Russell Smith Collection)

The foregoing notwithstanding, a total of more than 6,500 Spitfire Mk.Vs were built whilst a further 180 were converted from Spitfire Mk.I and II airframes. Service with the RAF continued until September 1945 when the marque was declared obsolete. In its latter years the Spitfire Mk.V in all its versions had been gainfully employed in secondary roles such as ground and pilot instruction. The type finally disappeared when the few remaining examples were tendered for scrapping in March 1948.

The Royal Air Force was not the only operator of the Spitfire Mk.V with examples being operated by the USAAF, in both British and American markings, plus the air forces of Egypt, Italy, Portugal, Turkey, and the USSR whose air force received 143 examples via the port of Abadan in 1943.

Spitfire Mk.VB AB910 is still operated by the Battle of Britain Flight although it is much changed from its operational days. Gone are the wing cannons and the original engine which has been replaced by that from a Mk.IX and associated four-blade propeller. (Big Bird Aviation Collection)

Whilst the Supermarine Spitfire went onto grab all the glamour and the glory, its counterpart, the Hawker Hurricane, found itself employed more and more on ground attack missions. (Nick Challoner)

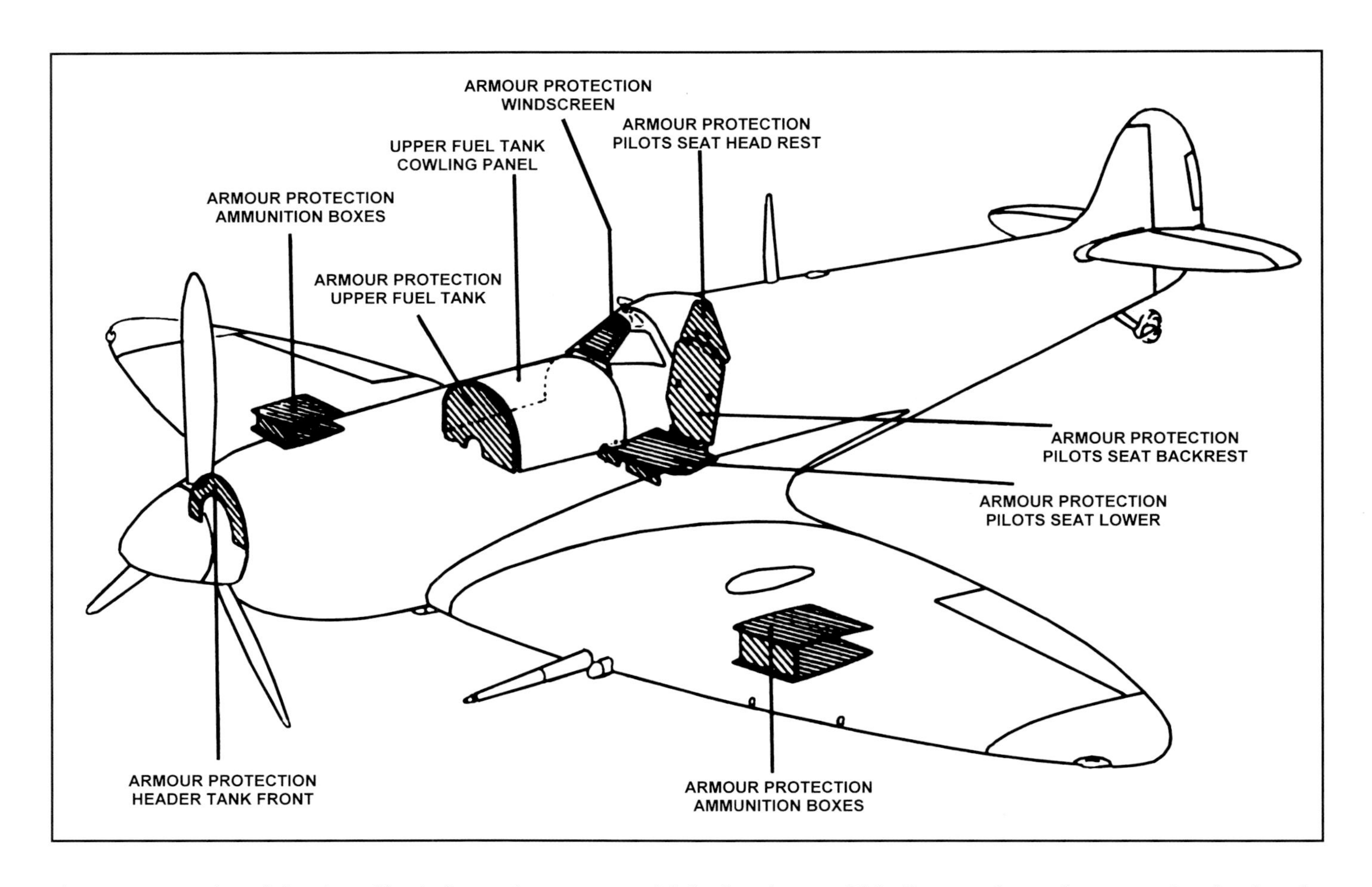

Armour protection of the aircraft's vital parts became essential during the war. This diagram shows that protection fitted to the Spitfire. (via Owen Morris)

Standing forlornly on its undercarriage is Spitfire Mk.VB(T) JK215. Very little is known of the operational history of this aircraft although in its stripped state it reveals much of what is normally hidden by external paneling. (C P Russell Smith Collection)

Constructed from the outset as a Spitfire Mk.VB(T) AB344 was flown off from HMS Eagle to Malta in March 1942. It was destroyed on the ground in the following month. This side-on view clearly shows the size and depth of the tropical filter unit first fitted to the Spitfire. (C P Russell Smith Collection)

Photographed in Germany on 29 April 1942 this aircraft was reported missing in action two days earlier. Built as a Mk.VB, AA940, was allocated to No. 303 Sqdn, a Polish-manned unit. Close inspection of the wing leading edges reveals that the guns have been fired as the protective fabric covering has been punctured by bullets. (C P Russell Smith Collection)

With a total of 57 hours and 30 minutes of flying time Spitfire Mk.VB, EP200, of No. 185 Sqdn sits in the desert after crash landing. Such was the force of the impact that the engine has been displaced from its Frame 5 mountings. This was to result in the aircraft being declared a write off. (C P Russell Smith Collection)

The port side of Spitfire Mk.VC AR501 reveals the gas warning patch located above the "A" of the squadron codes. (Dave Stewart)

With its gear still down AR501 flies past the camera. Still extant with the Shuttleworth trust this aircraft exhibits all the changes needed to create this version including the cannon armament and clipped wings. (Dave Stewart)

Wearing the MN codes of No. 350 Sqdn this Spitfire Mk.VB sits on the ground being guarded by a German soldier. Prominent on the pilot's head protector plate is the head rest developed in the light of combat experience. (C P Russell Smith Collection)

One of the countries that received Spitfires after the war was Portugal. Its Mk.VB, No. 19, is pictured here. (Eric B Morgan Collection)

The Spitfire was delivered to the Russian Air Force in large quantities. This is Mk.VC EP495 being prepared for its delivery flight. (Big Bird Aviation Collection)

This unidentified Spitfire Mk.VB served with the 334th Pursuit Squadron of the USAAF. (Eric B Morgan Collection)

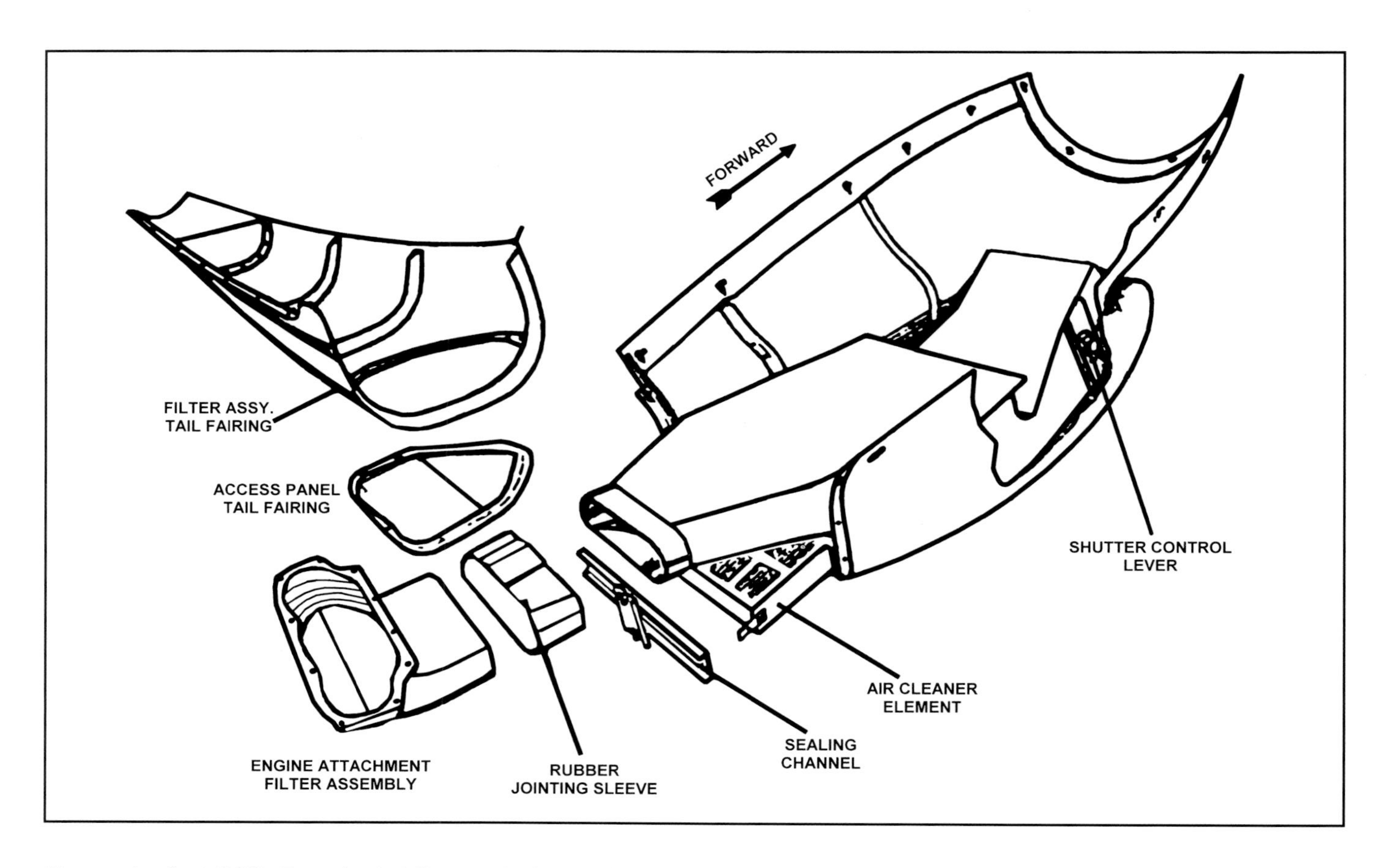

For use in the Middle East the Spitfire required extra filtering to keep the amount of sand particles to the minimum and thus reduce damage to the engine. (Chris Michell)

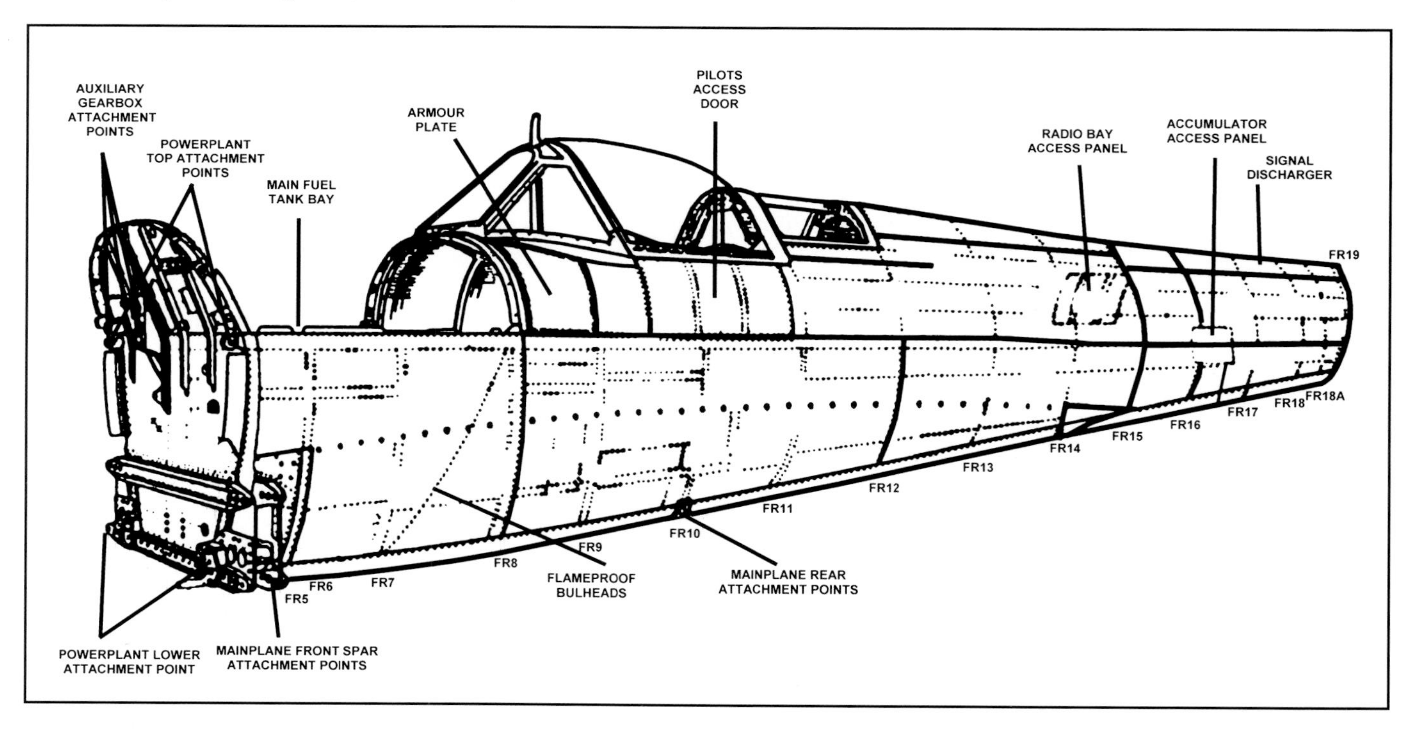

The layout of the skin plating remained the same throughout the life of the Spitfire, the only variations occurred in thickness. (Chris Michell)

5 Pink and Blue

Spitfires for PR and Speed

Given that the Spitfire was a small, speedy, and manoeuvrable aircraft it is hardly surprising that from its earliest days examples were drafted into units for reconnaissance duties. Prior to the appearance of the Supermarine product, photo reconnaissance was the remit of hastily converted bombers such as the Blenhiem.

The concept of a single aircraft undertaking high-speed photo runs over a specified target and departing in the same manner was a whole new idea when war was declared in 1939. It owes much of its inception to the efforts of one man, Flying Officer Maurice Longbottom. His experience in the field began prior to the fighting in Europe, when on behalf of the British Intelligence Service, he flew reconnaissance over the Italian fortifications in Libya in a modified Lockheed 12. First inklings of the concept were presented by Longbottom to the Air Ministry in August 1939 in a memo entitled "Photographic Reconnaissance of Enemy Territory in War". Part of the text covered in depth the methods to be used to penetrate enemy defences and avoid the attentions of anti-aircraft guns and enemy fighters. The aircraft defined in the memo was determined to be a single-seat machine that was stripped of extraneous weight, such as guns and radios, but had increased fuel tankage to extend its range. It also was capable of high speed and great manoeuvrability. It was recommended that it had an excess of power on take off and the ability to reach high altitudes.

It was fairly obvious from the memo that Longbottom was espousing the Spitfire for such a role, although modified from the fighter version. Much concentrated upon replacing the equipment required for the type's role as a fighter with fuel tanks. This would give a fuel system total three times greater than that of the normal aircraft at 240 gallons. In return, the range was increased to 1,500 miles at a cruising speed of 300 mph or 1,800 miles at 250 mph all at an altitude of 30,000 feet.

A dedicated photo reconnaissance organisation came into being just after war was declared. Placed in charge was Sidney Cotton who had planned Longbottom's flights over Libya. Given the rank of Wing Commander he set up his operation at Heston which was innocuously titled the Heston Flight. Original equipment for the flight was a handful of modified Bristol Blenhiem light bomber aircraft. However, it was obvious from the outset that no matter how carefully these aircraft were prepared they would be no match for a determined German air defence.

It took all of a month for Sidney Cotton to acquire the services of a pair of Spitfires from a reluctant Fighter Command. Serialled, N3069 and N3071, the modifications allowed on these two airframes

Originally built as a Mk.I fighter, L1004, acted as the prototype for the PR.XIII. A total of 25 were produced by converting some Spitfire Mk.Vs. Although fairly competent machines, pilots normally preferred the more dedicated reconnaissance aircraft. (C P Russell Smith Collection)

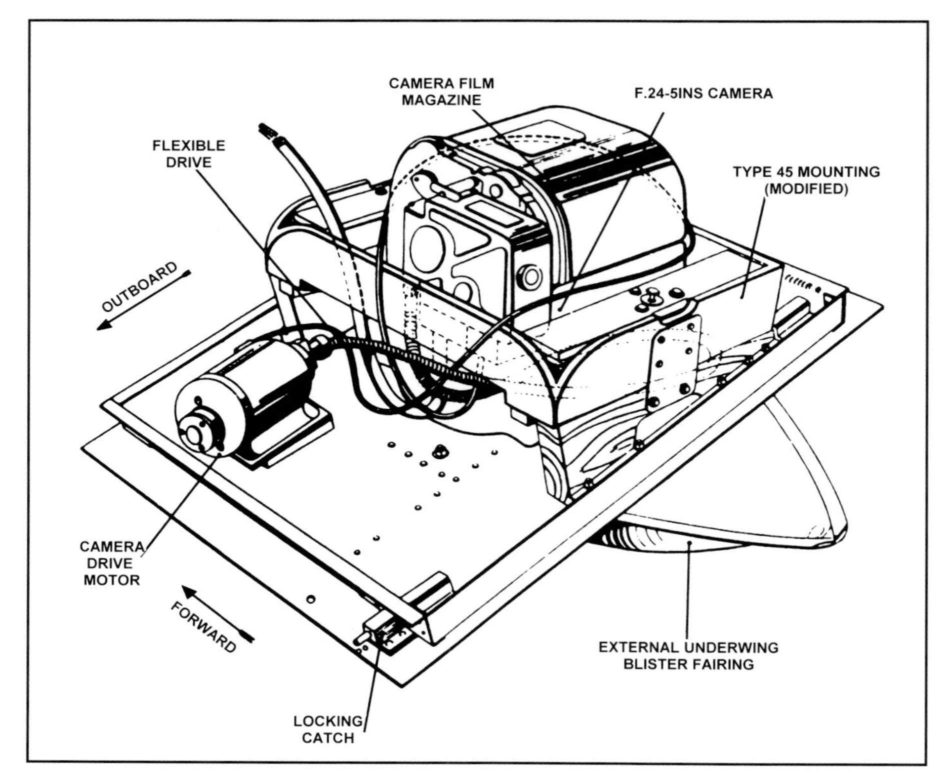

The earliest versions of the PR Spitfire had their cameras installed within the wing as Fighter Command wished to keep modifications to the airframe to a minimum until the type had proved itself in active service. (Big Bird Aviation Collection)

were kept to the minimum as further testing was deemed necessary. Thus, the camera installations were fitted in the wings in place of the original guns. No extra fuel tankage was authorised in these two airframes, by now designated PR. Mk.I, as the possible alterations to the aircraft's trim and handling needed investigation by the RAE.

The removal of the armament required that the gun ports be blocked off. As this was being accomplished the remainder of the airframe was having all its skin joints filled so that the airframe was as smooth as possible. The final crowning glory was a coat of light duck egg green, known as Camotint, that was deemed to be the right colour to blend in with the background of the sky at high altitude. To finally finish the aircraft, each was highly polished which gave the PR Spitfires an extra 15 mph above the top speed of their fighter siblings.

As it was becoming obvious that the Heston Flight had a specific purpose, it was decided to retitle it No. 2 Camouflage Unit which was as deceptive as its original nomenclature. This change took place at the beginning of November 1939 just in time for one of the Spitfires and the newly promoted Longbottom, now a Flight Lieutenant, to be detached to Lille in France as the Special Survey Flight. First operational sortie by a reconnaissance Spitfire was carried out on 18 November when a flight over the German city of Aachen was undertaken. Departing the French airfield of Seclin the mission was flown at an altitude of 33,000 feet, although navigation from this height was found to be difficult as study of the resulting photographs showed. Improvements in planning and increased precision navigation meant that the mission flown a few days later was accurate and delivered photographic intelligence of worth.

Having proved that the Spitfire was a viable reconnaissance platform, further airframes were transferred to the unit. However, a major change to the cameras was undertaken as the original short focal length cameras were only capable of defining large objects on the ground. Concentrations of troops and tanks were

One of the nations to acquire the Spitfire PR Mk.XI after the war was that of Norway. This example coded ZA-A served with No. 717 Sqdn based at Sola. This unit was later renamed No. 1 Photographic Reconnaissance Wing based at Garder moen. (C P Russell Smith Collection)

missed even under the most optimal conditions. The replacement cameras were Type F.24 with eight-inch focal length lenses which gave a higher degree of definition. Although, troops and tanks were still not visible from high altitudes

One further change was the installation of a fuel tank in the rear fuselage instead of the fighter's normal ballast weight. This extra 29 gallons resulted in the new version being designated the PR Mk.B. Also changed at this time was the camouflage finish, the original Camotint was replaced by a colour that later became known as PRU Blue. Even as the first of the new variants was being collected, the unit was undergoing a designation change again to the Photographic Development Unit.

The drive to improve the range of the reconnaissance Spitfire was to result in the appearance of the PR Mk.C which had a fixed underwing blister tank containing 30 gallons under the port mainplane. To counterbalance this protuberance another blister under the starboard wing contained a pair of Type F.24 cameras with eight-inch focal length lenses. Now that the PDU had a long-range Spitfire, it was time for the unit to carry out missions over Germany. The first of these was undertaken on 7 April 1940 over the major port of Kiel.

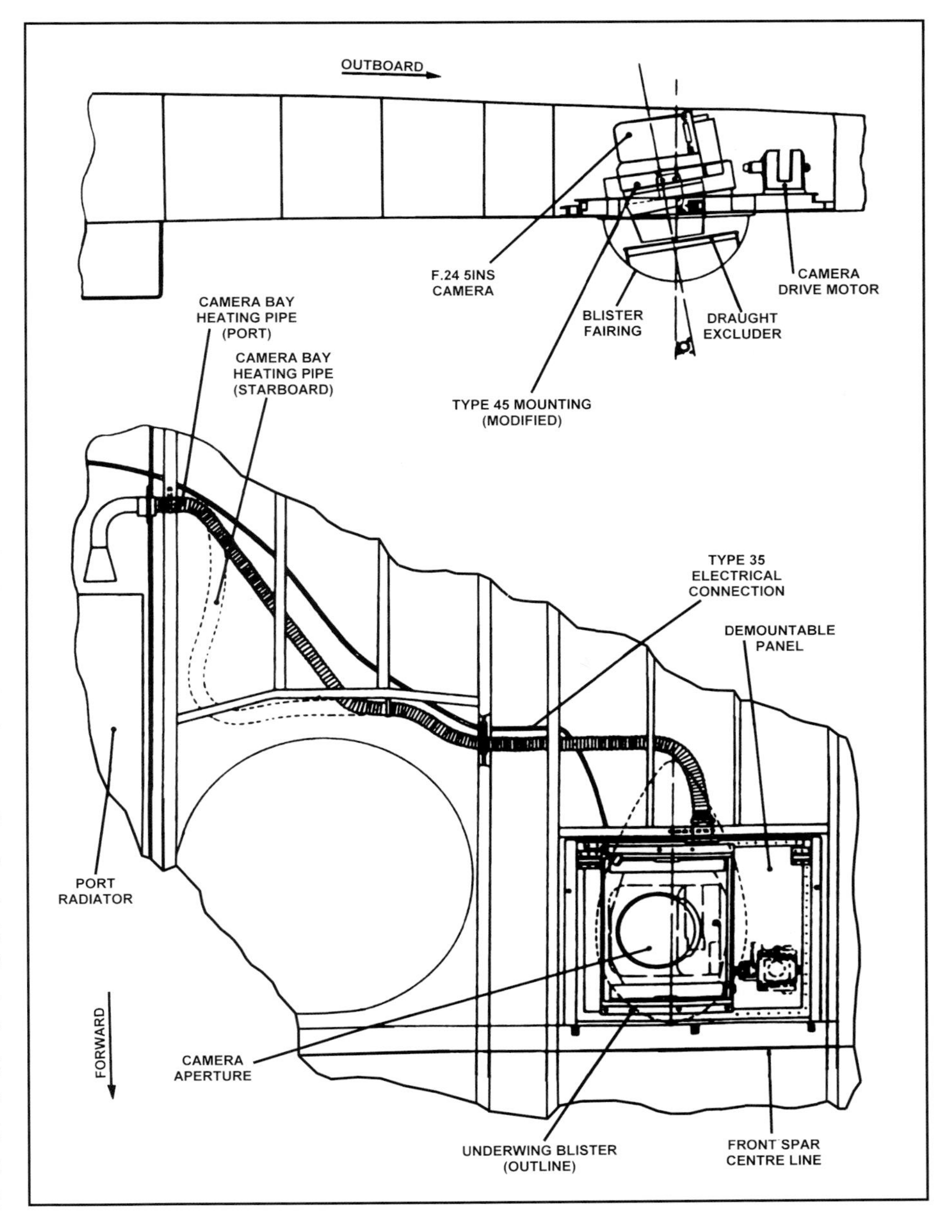

To keep the cameras warm during high-altitude flying and thus improve their efficiency, the wing gun heating system was modified to suit. (Big Bird Aviation Collection)

Having proved that the Spitfire could reach Germany and carry out its allotted task, the decision was taken to extend the type's range even further. Given that placing fuel in the rear fuselage would play havoc with the C of G range, attention turned to the one unused space in the airframe — the wing leading edge. Development work on the new type began in early 1940 when the PDU in consultation with Supermarine took over an aircraft on the production line and started the modifications required to produce the long-range reconnaissance Spitfire. The greatest change was the relocation of the cameras from the wings to a mounting unit in the rear fuselage. Having removed the operational equipment from the mainplanes they in turn were modified to accept integral fuel tankage in the leading edges.

After the conversion, a total of 118 gallons of fuel could be housed in the D-shaped integral tanks whilst a further 29 gallons was housed in a fuel tank behind the pilot. In view of the distances to be traveled, an extra 14 gallons of oil was housed in an overload tank in the port wing gun

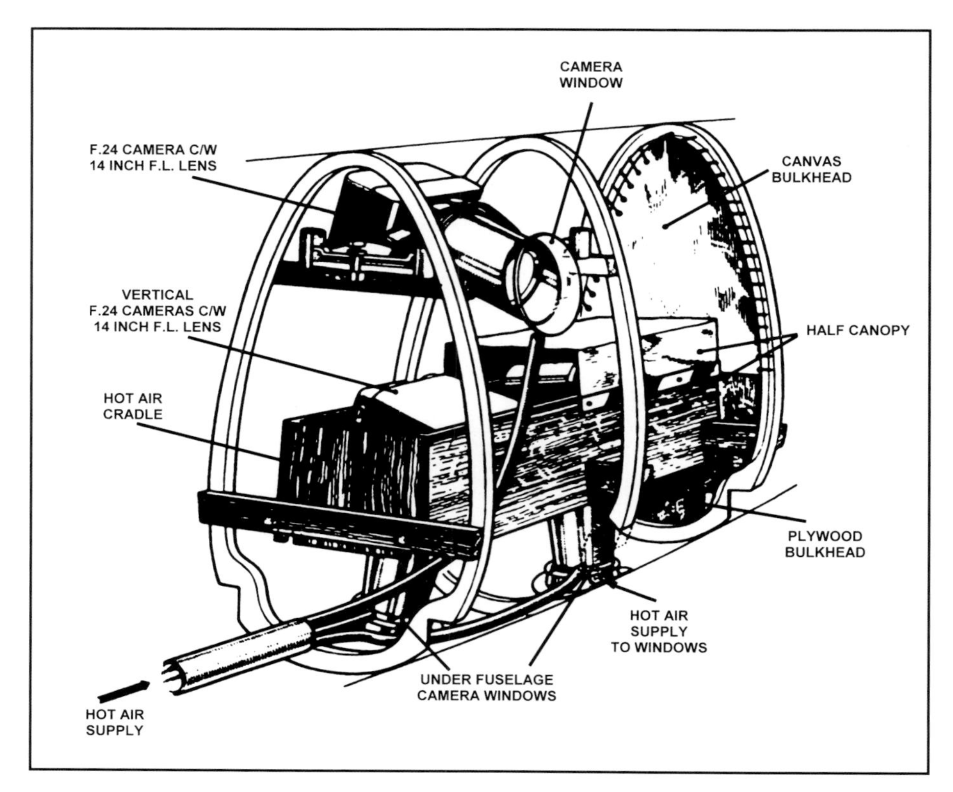

As confidence in the Spitfire as a PR platform developed so there was a reshuffle of resources within the airframe. Fuel mainly gravitated to the wings whilst the camera installation became ensconsed within the fuselage. (Big Bird Aviation Collection)

bay. The camera installation comprised of a pair of F.24's with either 8- or 20-inch focal lengths or a pair of F.8 cameras with 20-inch lenses. Performance especially during the take off phase was described as difficult as the C of G limit was so far aft.

This problem, however, did not stop the first sortie from taking place on 29 October 1940 when a PRU pilot flew to Stettin. The PDU, upon transfer to Coastal Command for administrative purposes, had been renamed the Photographic Reconnaissance Unit in June 1940 which further defined its role. Following on from this first mission, which lasted 5 hours and 20 minutes, further sorties were successfully flown over Marseilles, Trondheim, and Toulon.

Following on from the first two hand-built conversions the next Spitfire PR.Ds to emerge came from the production line. These airframes dif-

This official portrait of a Spitfire PR Mk.X quite clearly shows its antecedents from the Spitfire Mk.VIII fighter. In fact the only obvious changes from the fighter, apart from the paint finish, are the extra intake on the port side of the nose and the deepened oil tank. (C P Russell Smith Collection)

One of only five Spitfire floatplanes built, this is Mk.V W3760 which acted as the prototype for this small batch. (Eric B Morgan Collection)

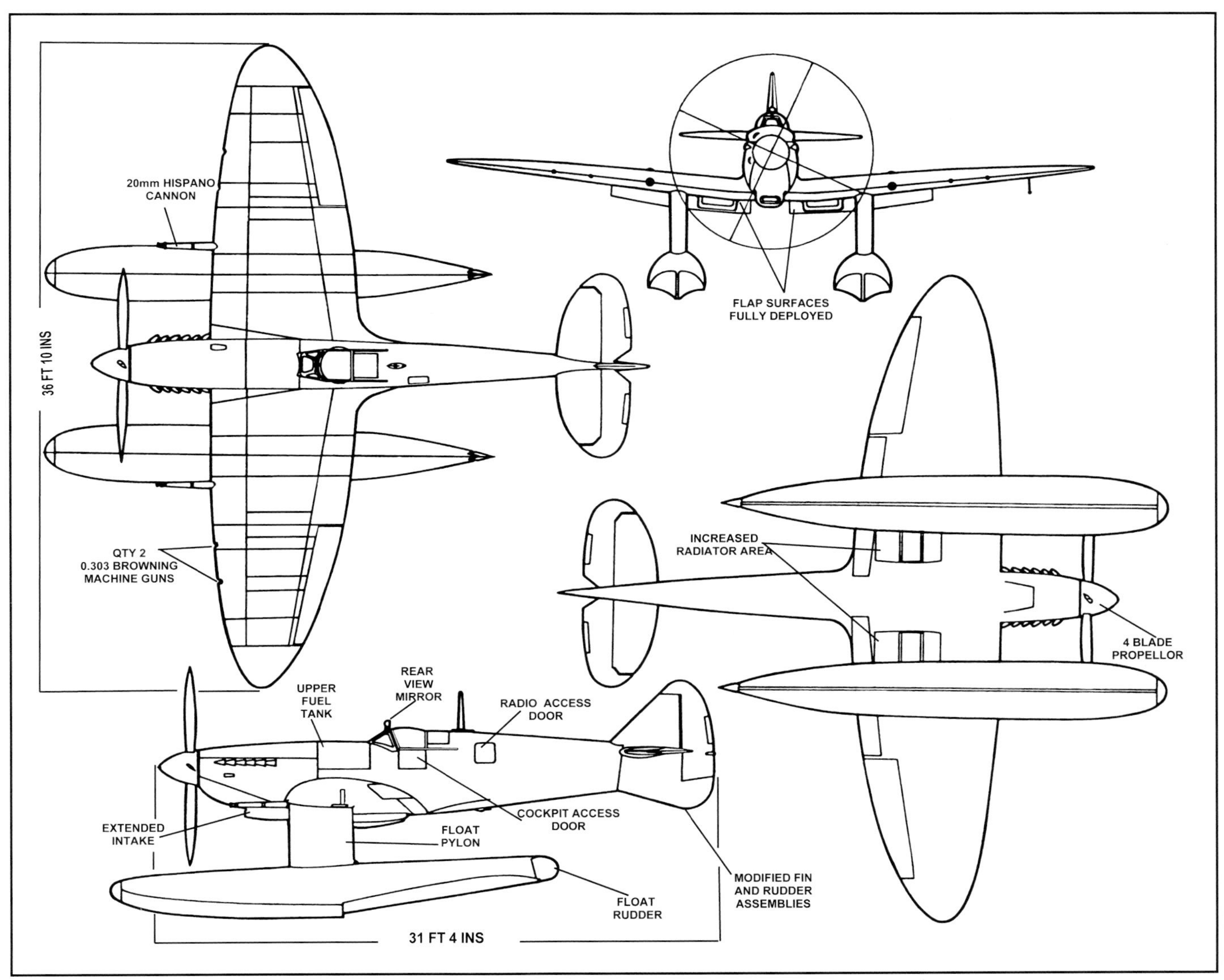

Reminiscent of earlier days and the Spitfire floatplanes. Only a handful were converted and this is the general arrangement for the Mk.IX version. (Big Bird Aviation Collection)

fered from the originals in that the rear fuselage tank was deleted and the fuel transferred to larger tanks in the wing leading edges, which grew to accommodate 66 gallons. The original powerplant, the Merlin III, was replaced by the more powerful Merlin 45 engine from which a tapping was taken to provide a cabin heating system. Also reinstated in this version was the radio system. A final total of 229 Spitfire PR Mk.C, later designated the PR Mk.IV, were built for the RAF during the period 1941 to 1942.

Further versions of the PR Mk.I were built although these only appeared in small quantities. The version produced immediately after the "D" was not surprisingly designated the PR Mk.E. There was only one aircraft built to this specification and it was intended to try out low-level oblique photography. This aircraft, N3117, carried an F.24 camera in a bulged mounting under each wing. Running in at low level the cameras were directed out at right angles to the direction of flight and were canted downwards at an angle of 15 degrees. The first such mission was flown in July 1940 and was deemed very successful as it was discovered that oblique photography could produce results under conditions that eluded vertical camera sorties.

Although designated the PR Mk.F this version of the reconnaissance Spitfire actually appeared before the long-range "D". This variant was seen as the "super long-range" Spitfire and featured a blister fuel tank under each wing which contained 30 gallons of fuel and was backed up by the rear fuselage tank behind the pilot's seat containing a further 29 gallons. The camera installation consisted of a pair of F.24 cameras with 8-inch focal lenses, these later being upgraded to 20-inch lenses. Further modifications later in the type's service would allow the installation of a single F.24 with a 14-inch focal length lens mounted obliquely. Operational service of the PR Mk.F began in July 1940 and allowed photography up to a range of 100 miles. Once the Mk.F was established in service use, the remaining Type B and C aircraft were converted to the new standard.

Only one other version of the early-build reconnaissance Spitfires was to be manufactured. This was designated the PR Mk.G. Very much a fighter reconnaissance aircraft, this version retained the eight machine gun armament for self defence plus the gunsight and laminated reinforced front windscreen. Fuel tankage consisted of the normal fighter contents plus the overload tank containing 29 gallons behind the pilot. Designed

Spitfire Mk.VI, AB200, was originally ordered as a Mk.V. It spent much of its time as a trials aircraft first being used for Merlin 45 trials. Pictured here it has the extended long span wing, the original pair being buckled during a high-speed dive. (C P Russell Smith Collection)

for oblique reconnaissance sorties, hence the retained armament, the camera installation consisted of an F.24 camera complete with a lens of 14-inches focal length. This could be directed to port or starboard as required and was supported by a pair of vertically mounted F.24 cameras, one with a 5-inch length lens and one with a focal length of 14 inches The range of cameras installed meant that the Spitfire could cover altitudes between 2,000 and 10,000 feet quite comfortably. As if to emphasise their unique role, most of this version of reconnaissance Spitfire were painted a pale pink overall. In service use the PR Mk.G, later the PR.Mk.VII, was operated by No. 1416 flight which was to later evolve into No. 140 Sqdn.

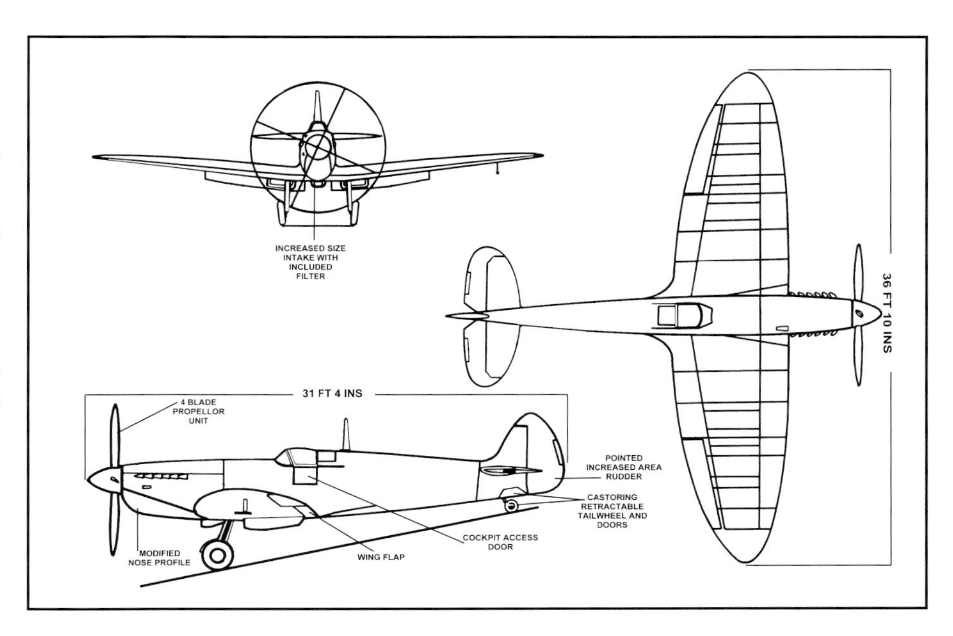

This general arrangement diagram covers the modifications essential to create the PR Mk.X and Mk.XI Spitfires. (Big Bird Aviation Collection)

During their service career the alphabetic Spitfires were operated by No. 1 PRU, operating from the UK, No. 2 PRU flying from Oakington in the UK, and No. 3 PRU which operated out of Heliopolis, Egypt. Numbered operating units included Nos. 541, 542, and 543 Sqdns amongst others.

The first tranche of reconnaissance Spitfires owed their origins to the Mk.I and II fighters. Later versions were to owe their lineage to the much produced F.Mk.IX. Three basic versions were built, these being designated Mk.X, Mk.XI, and Mk. XIII respectively.

The first of these, the Mk.XI, had been preceded by a quick conversion of some fighter Mk.IXs for squadron service. The Mk.XI was an amalgam of the wings from the PR Mk.D, the Merlin 61 powerplant from the fighters, all added to a modified fuselage developed from that of the Mk.VII and VIII. First production examples began to leave the Supermarine facility in November 1942 and replaced the earlier PR. Mk. IV in operational squadrons. Earlier versions of the reconnaissance Spitfire had revealed that during tactical missions the normal long focus lenses missed a great deal. To improve the available low-level coverage, a two F.8 cameras were installed in blisters under the wings where they were canted to give maximum coverage. The type was also used to test the maximum speed possible with the Spitfire which was established at 606 mph, although the engine exploded in flight and shed

Complete with SEAC markings Spitfire PR.Mk.XI, PL951, has suffered a slight landing mishap which resulted in it ending up on its nose. After recovery work the aircraft returned to service and served until 1947. (C P Russell Smith Collection)

Although a slightly indistinct portrait this is indeed a rare view of a Spitfire PR.1F belonging to the PDU whose pink camouflage it sports. The blisters under the wings are external tanks that contain an extra 30 gallons of fuel each. The aircraft also sports an enlarged oil tank under the nose. The cameras were housed in the rear fuselage. (Eric B Morgan Collection)

the propeller. The pilot eventually brought the aircraft back for a dead stick landing. Notwithstanding this attempt at self destruction, a total of 471 Spitfire PR Mk.XI were manufactured up to 1944 from normal Spitfire Mk.IX fighters.

As if to prove that the nomenclature system was eccentric to say the least, the next version of the reconnaissance Spitfire was designated the PR Mk.X. Very similar to the PR.XI the main differences included a pressure cabin and a Lobelle sliding hood. In the end, only 16 of this version were manufactured and were operated by Nos. 541 and 542 Sqdns in the high-altitude reconnaissance role.

The final Merlin reconnaissance Spitfire was designated the PR Mk.XIII. A total of 26 of this version were converted from both fighter and reconnaissance versions, in particular the Mk. II, Mk.V, and PR Mk.G. The prime purpose of these aircraft was to provide photo reconnaissance at low level over the Normandy beachhead during the D-Day landings. Operated by Nos. 4, 400, 541, and 542 Sqdns the Mk.XII differed from other variants in toting four Browning machine guns for self defence. Its powerplant was a specially derated Merlin 32 optimised for high speed, low altitude work whilst the camera installation was that of the PR Mk.G slightly modified.

Foreign operators of the reconnaissance Spitfire after the war included the Royal Norwegian Air Force which operated three under the aegis of No. 1 PR Wing based at Gardermoen. One unusual resale was purchased for aerial survey and mapping work in Argentina. This Spitfire PR.XI, registered LV-NMZ, routed to its new owners via Dakar in Senegal and Natal in Brazil. To ensure that it could reach its destination, the aircraft not only had all its tanks fully replenished but it also had a ventral tank containing 170 gallons under the fuselage. The Spitfire flew the complete journey on 5 May 1947 and covered 1,850 miles in eight hours.

Developments in both the fighter and reconnaissance field were to lead to the Spitfire F.VI, which was intended to replace the stopgap Mk.V in the high-altitude interceptor role. In order to make this a reality, a pressure cabin was installed in airframe X4942 and test flown on 5 July 1941. Overall the flight was a success, although the pilot did report that a fine oil mist from the compressor did enter the cabin and caused an unpleasant smell. Another report cited a slight feeling of restriction upon the flying controls that was caused by the rubber seals. Further comment also pointed out the lack of temperature control which resulted in an excess of heat at high altitudes. Unlike the normal fighters, the original slide-back canopy was replaced by a clamp-on type which was seen as an interim especially in view of the problems presented by escape.

Although Supermarine had pressed ahead with developing a cabin pressurisation system, it was not alone in developing such a unit as the RAE had been working along similar lines. To test the pressure cabin from the RAE, Spitfire Mk.V, R7120, had this system installed. The resultant flight tests were to prove that the RAE unit was effective in many areas, thus the production version of the aircraft would include the best features from both systems.

Production versions of the Mk.VI were built without the side entry door normal to the fighter and were fitted with a rerated Merlin 61 engine more suited to high altitudes which drove a four-bladed propeller. No special strengthening was undertaken to the fuselage structure as the system installed was of the partial pressurisation type with a differential of 2 psi. This required that the pilot would still need to wear an oxygen mask as the aircraft could only simulate a height of 28,000 feet at an altitude of 37,000 feet. Although the test pilot had recommended that the canopy be of the sliding type, first production aircraft were manufactured with the clamp-on type, this in turn needing an inflatable canopy seal to reduce leakage from that zone. Later aircraft had the Lobelle type of sliding hood fitted from new which had a powerful ejection spring attached for emergency escape purposes. To improve handling at increased altitudes the wingspan was increased to 40 feet 2 inches.

The first aircraft was rolled out in September 1941 with the second moving to A&AEE Boscombe Down in February 1942 for testing and evaluation. Operational service began in April when No. 616 Sqdn based at Kings Cliffe received its first examples. Further aircraft were used by the AFDU based at Duxford which gave a full evaluation of the type. The report from this body stated that the Spitfire Mk.VI behaved in a similar manner to the Mk.V, however there were improvements in its performance above 30,000 feet. There were some minor problems reported with the cabin heating at low altitudes and a few leaks through the various bulkhead penetrations. A total of 100 aircraft are reported to have been delivered by Supermarine although only 97 were accepted according to RAF records.

Service use by the RAF included operation by Nos. 91, 310, 313, and 504 Sqdns before the type was transferred to second line units in late 1943.

Following on from the first pressurised aircraft came the Spitfire Mk.VII which was a more powerful and strengthened aircraft. Changes from the previous aircraft included a retractable tail wheel, a 14-gallon fuel tank in each wing leading edge, rectangular-shaped radiators under the wings, and a Lobelle-type canopy suitable for pressure cabin use. To improve performance at high altitude the wing tips were extended allied to which the ailerons were shortened to improve the rate of roll and reduce potential flutter. Subject to a low volume rate of production, the first Spitfire Mk.VII rolled out in August 1942 with deliveries following on to the first operational unit, No. 124 Sqdn. Eventually a total of six squadrons flew the type mainly in the high-altitude fighter protection role for which purposes it retained full armament and armour protection, a contrast to the original premise of flying with reduced weaponry and removed armour

This excellent side-on view of a Spitfire Mk.VII, BS142, clearly shows the extended wings fitted to this version. Fitted with cannon armament this aircraft also sports the intake on the nose for the cabin pressurisation system. (C P Russell Smith Collection)

Spitfire HF.Mk.VII, AB450, was converted from a Mk.V to act as the prototype for this marque. (Eric B Morgan Collection)

which greatly enhanced its high-altitude performance. A final total of 141 aircraft were taken on charge by the RAF and issued to Nos. 118, 124 , 131, 312, 435, and 616 Squadrons.

After the Mk.VII's frontline was completed some aircraft were employed for meteorological duties by No. 1402 Met Flight in which role they continued until replaced by the Spitfire PR.19. Others of the type were employed on booster trials during late 1945 at Farnborough at the behest of the RAE. Two airframes were converted by Heston Aircraft and featured LOX boosting to the installed Merlin 71 engine which increased its output to 300 bhp. In real terms this pushed the performance at 40,000 feet up by 27 mph and even more substantial improvements were available above 41,000 feet.

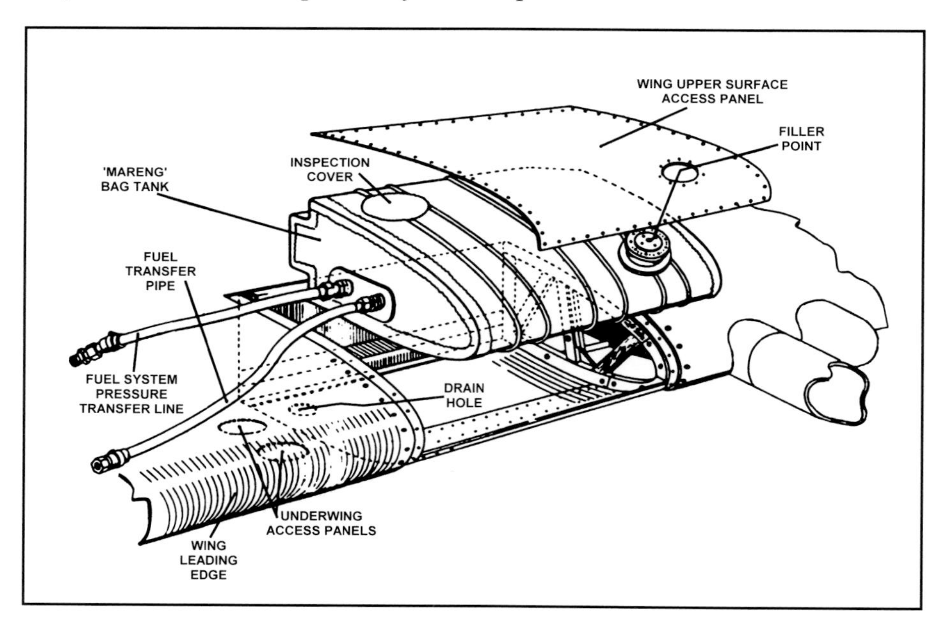

To improve the range of the Spitfire Mk.VIII an extra fuel tank was installed outboard of the cannon mounts. (Big Bird Aviation Collection)

From the outset of Spitfire development the type was always subject to interesting modifications, some more esoteric than others. The first of these was euphemistically called the Speed Spitfire with which it was intended to capture the world speed record for single-engine, piston-powered propeller aircraft. To this end Mk.I, K9834, was removed at an early stage from the production line for conversion.

Streamlining and strengthening were the areas most concentrated upon. One of the first zones undergoing such treatment was the main engine bulkhead at frame 5. Here the airframe had to be strong enough to absorb the torque output of the Merlin sprint engine which in turn drove a coarse pitch four-blade wooden propeller. Overall the airframe was flush riveted with all gaps being filled and rubbed down before being finished in a scheme of royal blue on the upper surfaces with silver underneath and a fuselage flash in a similar colour.

To improve the range of the Spitfire Mk.VIII an extra fuel tank was installed The Spitfire PR Mk.XI was initially developed from the Mk.VIII although most were later converted from the Mk.IX fighter. This is one of the earlier series powered by a Merlin 61. (C P Russell Smith Collection)

First flights of the aircraft, serialled N.17, began in February 1939 from Eastleigh during which a maximum speed of 408 mph at 3,000 feet was achieved. Problems with the cooling system and an increase in the world record speed by the Heinkel 100 meant the Speed Spitfire attempt was delayed. Even as the aircraft was being repaired, events in Europe overtook the attempt on the speed record and it was to be canceled. Although the aircraft was the fastest in the RAF inventory the differences between it and the standard variants meant that it could not be converted back to a fighter, therefore it was delivered to the Heston flight. However, the lack of fuel in this airframe equated to a lack of range. Thus, the Speed Spitfire ended its days as a unit hack, a duty it performed until 1946.

Possibly the most unusual version of the Spitfire was that known as the Aerolite Spitfire. This was an attempt to develop and manufacture an aircraft built of various resins. The reasoning behind this most unusual development was a concern that the war may cause a shortage in the bauxite ore used to produce Aluminum. A contract to produce an Aerolite Spitfire was issued in February 1941 for one fuselage. Using an idea in use today, that of cross laying the fibres and curing a resin mix, some layers of material were produced and riveted to the test fuselage frames. Although only two fuselages were produced, the research generated from them led to panels being manufactured for operational aircraft and a whole range of adhesives being developed suitable for the aviation industry. The final compliment to the fibre-based type panel came with the development of carbon fibre autoclaved aircraft components.

Although the plastic Spitfire remained no more than a developmental novelty, there was one serious attempt to extend the versatility and range of the type. The Invasion of Norway in April 1940 by the Axis forces and the realisation that there were not enough usable airfields to repulse them caused designers to attempt to extend the Spitfire's versatility and range.

Initial thoughts centred about fitting the complete float set from the Blackburn Roc turret fighter of the FAA underneath both the Spitfire and the Hurricane. Although these were larger than necessary, the flight trials using Spitfire Mk.I, R6722, found that the performance on water was

The Spitfire PR Mk.IV was originally known as the PR.1D. Unlike the fighters, this type featured a rounded windscreen, bulged canopy, and camera ports under the fuselage. (C P Russell Smith Collection)

A later photograph of the Spitfire Mk.III prototype reveals that it has been fitted with a pair of short span wings. (C P Russell Smith Collection)

without problem. However, as the floats were oversized, they reduced the Spitfire's top speed by 10 mph. A further attempt at perfecting the float Spitfire was made using a Mk.V, W3760, fitted with a specially designed set of floats. Improved longitudinal directional control was required which was catered for by fitting a rudder extension that protruded below the fuselage line.

Subsequently another two aircraft were converted to the same standard and ambitious plans drawn up for their use, mainly centred upon the Greek Islands. Eventually this plan came to naught and the aircraft were employed for training purposes upon the Great Lakes in Canada. One final Spitfire, a Mk.IX serialled MJ982, was also converted to floatplane status for potential service in the Far East war. Although thoroughly tested throughout 1944, the eventual requirement for such an aircraft was dropped and the Spitfire converted back to standard fighter status.

Destined never to attempt the record it was built for is the Speed Spitfire, N17, on show at the International Aeronautical Salon in Brussels during July 1939. Modified beyond restoration to standard, the aircraft ended its days as a unit hack with the PRU. (Eric B Morgan Collection)

Spitfires in Colour

Royal Air Force Colours

The prototype Spitfire prior to painting in its blue-green finish first flew in an unpainted state. This was to be the only time that the type flew as such until hostilities finished in 1945. As the clouds of war gathered over Europe the Air Ministry decreed that all frontline aircraft should be camouflaged. Thus, the Spitfire was clothed in a finish of dark earth and green on the upper surfaces whilst the underneath was decorated in light blue. During the Battle of Britain the undersurfaces were painted half black, half white in an effort to aid anti-aircraft gunners in differentiating between the aircraft of the Royal Air Force and the Luftwaffe. As the war progressed the light blue surfaces returned, but were augmented with a sky blue spinner and rear fuselage band. With the arrival of the Spitfire Mk.V the camouflage changed to green and dark gray on the upper surfaces being augmented by yellow leading edges to the wings. The appearance of the Mk.IX meant a change to the undersurface finish which became light gray. This was the scheme applied to the greater majority of the Spitfire Mk.IX production run although some dedicated to the Tac R role had medium sea gray applied to the upper surfaces. Extra splashes of colour applied to these combat aircraft included coloured spinners and the inevitable D-Day stripes on the fuselage and wings.

For Spitfires destined for service in the Middle East an upper surface finish of sand and earth was applied over undersurfaces of azure blue. Those aircraft, mainly Mk.VIIIs, were shipped overseas wearing a dark earth and green over gray scheme complemented by markings in which the red portion had been removed. This last was to ensure that the confusion with the Japanese rising sun markings was reduced.

The field of aerial reconnaissance was probably the most colourful during the war years as the camouflage experts worked their way through a green shade known as Camotint via a shade of pink to the more ubiquitous PRU blue. Some aircraft engaged in more close support reconnaissance missions wore standard camouflage, although these were few in number.

RW393 is a Spitfire LF.XVI that is preserved by the RAF. It is pictured here in standard postwar temperate colours complete with the codes of No. 603 Sqdn, an R Aux AF unit. (Big Bird Aviation Collection)

One of the most authentic Spitfire Mk.IIs in preservation is P7350 seen here approaching to land. Unlike other aircraft the flaps of the Spitfire were used for landing only, although some experiments were carried out using the flaps for improving stability during dive-bombing operations. (Nick Challoner)

With its wings removed this Spitfire Mk.IX sits on jacks and trestles whilst undergoing in-depth maintenance. Removal of the cowling panels has revealed the engine and its ancillary components. This port side view of the Merlin engine clearly shows the location of the oil and header tanks plus the support struts and exhaust manifolds (Chris Michell)

Contrary to first observations this aircraft, Spitfire IA AR213, looks more like a Mk.IX due to the installation of a late-model Merlin engine and four-blade propeller. The camouflage and markings are fairly accurate consisting of earth and dark green uppers with sky blue below. The codes, PR, are those of No. 609 Sqdn although AR213 only served with 57 OTU before withdrawal. (Big Bird Aviation Collection)

Spitfire Mk.VB, AB910, has a claim to fame as the aircraft that took a WAAF for a flight on the tailplane after the pilot forgot she was there acting as a counterbalance. Both Margaret Horton and the Spitfire survived. The finish is the late temperate complete with duck egg blue tailband. In RAF service the aircraft flew with No. 416 Sqdn and 57 OTU before retirement. The coding, MD, is that of No. 133 Sqdn. (Big Bird Aviation Collection)

Flying in formation are a Spitfire Mk.VIII and a Mk.XVI. The codes on the Mk.VIII are those of No. 145 Sqdn. (Danny Jacquemin)

Spitfire Mk.IX, MK732, is resplendent in full D-Day markings and the sky codes of No. 485 Sqdn. As this is a restored aircraft the striping is perfectly formed unlike many of those that participated in the actual operation where the accuracy was wavy to say the least. (Big Bird Aviation Collection)

Sporting the earlier rounded rudder fitted to the Spitfire Mk.IX is MH434. Whilst in service with No. 222 Sqdn it was credited with shooting down an FW-190. After use by the RAF this airframe flew with both the Dutch and Belgian Air Forces prior to entering private hands. With its flaps in the landing position MH434 wearing the markings of No. 315 Sqdn is moments from touchdown.
(Nick Challoner)

Currently owned and operated by the Shuttleworth Collection this Spitfire LF.VC, AR501, is one of the most authentic in preservation as it still wears the NN codes that it wore in service with No. 310 Sqdn. This particular unit was purely manned by exiles from Czechoslovakia whose roundel is visible below the windscreen. (Dave Stewart)

This beautiful shot from the rear of Spitfire AR501 reveals the strengthened wing root walkway and strengtheners on the wing upper surfaces. (Dave Stewart)

In contrast to the clipped-wing Spitfire Mk.VC, this is a Mk.VB with everything out and down preparing to land. Unlike the Mk.VC this aircraft has standard wingtips fitted. (Damien Burke)

Triple two Squadron are the codes that adorn this Spitfire Mk.IX, MH434, a privately-owned aircraft. Given the expertise available to rebuild and refurbish Spitfires it is not surprising that the current standard of engineering is high. (Big Bird Aviation Collection)

The markings that adorn this Spitfire Mk.XVI represent those of No. 41 Sqdn although they would have been more at home on the Griffon-powered Spitfire F.21. (Big Bird Aviation Collection)

Pink camouflage and D-Day stripes adorn this beautifully restored Spitfire PR.XI as it flies past the camera. (Dave Stewart)

The Spitfire Mk.VIII was flown extensively by the RAAF in the Far East. A58-758 is equipped with both cannon and machine guns. (Big Bird Aviation Collection)

This shot of AR501 shows the blunter spinner fitted to aircraft with a three-blade propeller. Unlike aircraft in wartime this Spitfire has had its enlarged head rest removed for comfort. (Dave Stewart)

The Spitfire Mk.VIII was intended to be the definitive article with its strengthened fuselage and retractable tail wheel and other modifications. Eventually a total just exceeding 1,000 were built although the airframe formed the basis of many other variants. This particular aircraft, MT719, never saw any active service with the RAF being passed onto the Indian Air Force for use. The markings are those of No. 17 Sqdn based in the Far East, note the lack of red in the roundel and tail flash. (C P Russell Smith Collection)

As the production of the Spitfire Mk.IX continued, the rudder was changed to that of a more pointed nature which with a slightly increased chord improved the longitudinal stability of the type. Wearing D-Day striping this is ML417 resplendent in the duck egg green codes of No. 443 Sqdn, a Canadian unit with which it served in 1944. (Nick Challoner)

For the Masses

6

The Spitfire Mk.VIII and Mk.IX

The Spitfire Mk.VIII evolved from the earlier Mk.III of which only one example was built. Inherent in the design was an increase in overall airframe strength, the intention of which was to cure some deficiencies and weaknesses in the original design. When asked to develop a new fighter for the RAF, Supermarine drew upon experience gained with the Mk.III and the high-altitude Spitfire Mk.VII. The resultant aircraft featured the same strengthened airframe, retractable tail wheel, wing leading edge fuel tanks, and the Merlin 60 series of engines.

Production began in late 1942. The first examples appeared in June 1943 when JF274 began its maker's flight trials taking over from the converted Mk.III, N3297, which had acted as the version prototype. This Mk.III continued in use as a developmental aircraft being utilised for a variety of engine trials. Early production Spitfire Mk.VIIIs were delivered with the extended wingtips reminiscent of those fitted to the Mk.VII high-altitude fighters. Eventually there were three versions of the Spitfire Mk.VIII built: the standard fighter "F" version; the HF optimised for escort work and high-altitude interception, complete with extended pointed wingtips; and the LF designed for operations at low level.

The first aircraft for the RAF, JF462, was delivered in April 1943 and was powered by a Rolls- Royce Merlin 66. This particular aircraft was initially employed for testing a modified undercarriage to resolve some of the discrepancies in behaviour experienced by normal units. Testing by A&AEE reported no adverse effects during ground manoeuvres.

Contrary to all normal perceptions, the early Spitfire Mk.XVIs were constructed with high-back fuselages. This immaculately restored example, TB597, is pictured at 5 MU Kemble prior to delivery to Paris for a museum. (Ray Deacon)

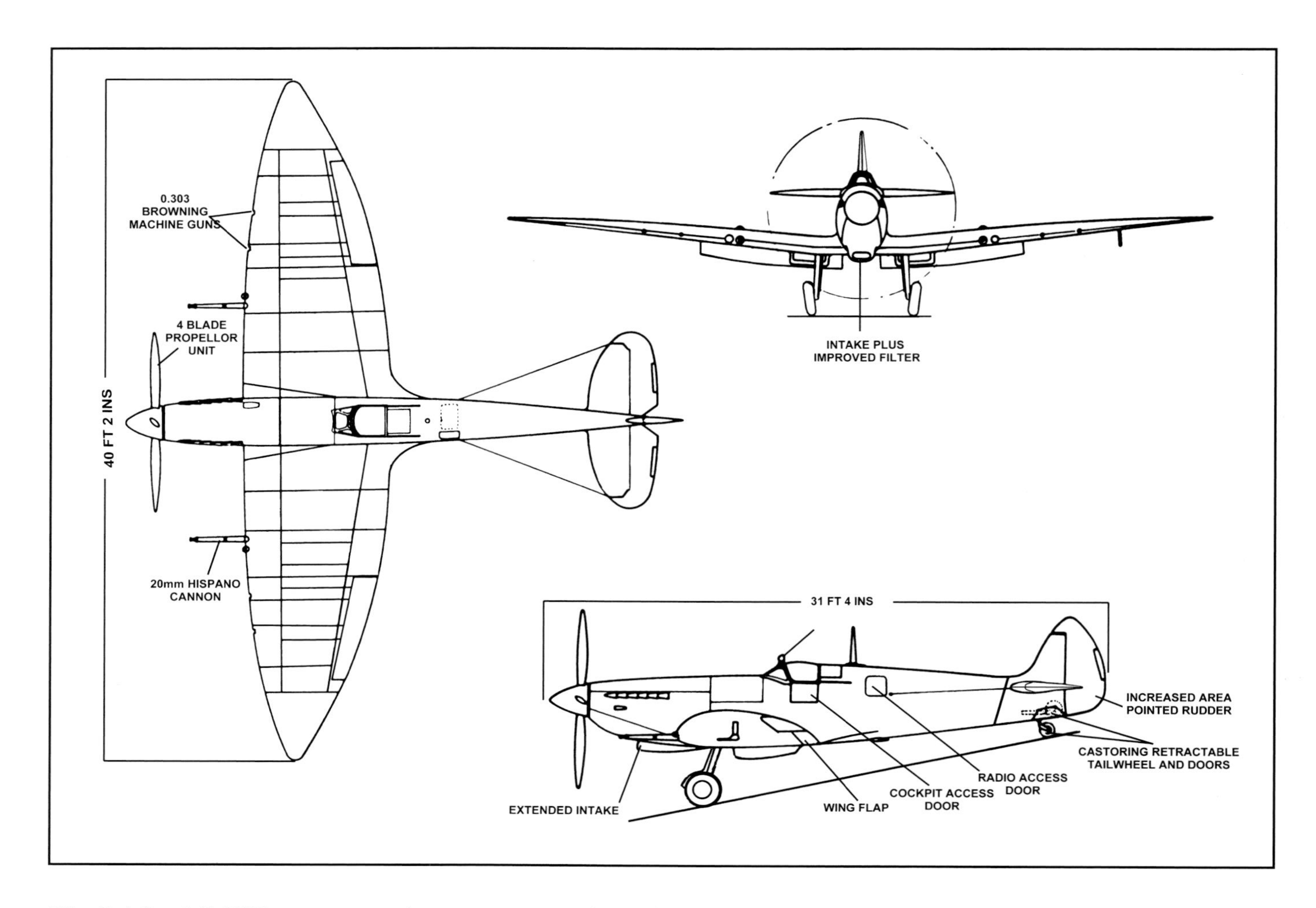

The Spitfire Mk.VIII was a great improvement on the earlier versions. Notable changes included a stronger airframe, retractable tail wheel, and Vokes air filter as standard. (Big Bird Aviation Collection)

This official view of Spitfire Mk.VIII, JF463, reveals an aircraft from one of the early batches complete with rounded rudder and Merlin 63 engine. (C P Russell Smith Collection)

As the new version of the Spitfire featured the Merlin 60 series of powerplants, the resultant installation required a slight extension to the engine cowlings. This meant that the Spitfire, already a difficult aircraft to land and manoeuvre on the ground, began to suffer a series of ground "dinks" as pilots touched down. To try and alleviate this problem, aircraft JG246 was despatched to Rotol Airscrews Ltd for investigation. A cropped propeller assembly was fitted during a series of trials to see how much of a reduction could be carried before the aircraft began to suffer performance loss. Eventually a length of 8 feet 3 inches was seen as the shortest for safe flight to an MU for repair.

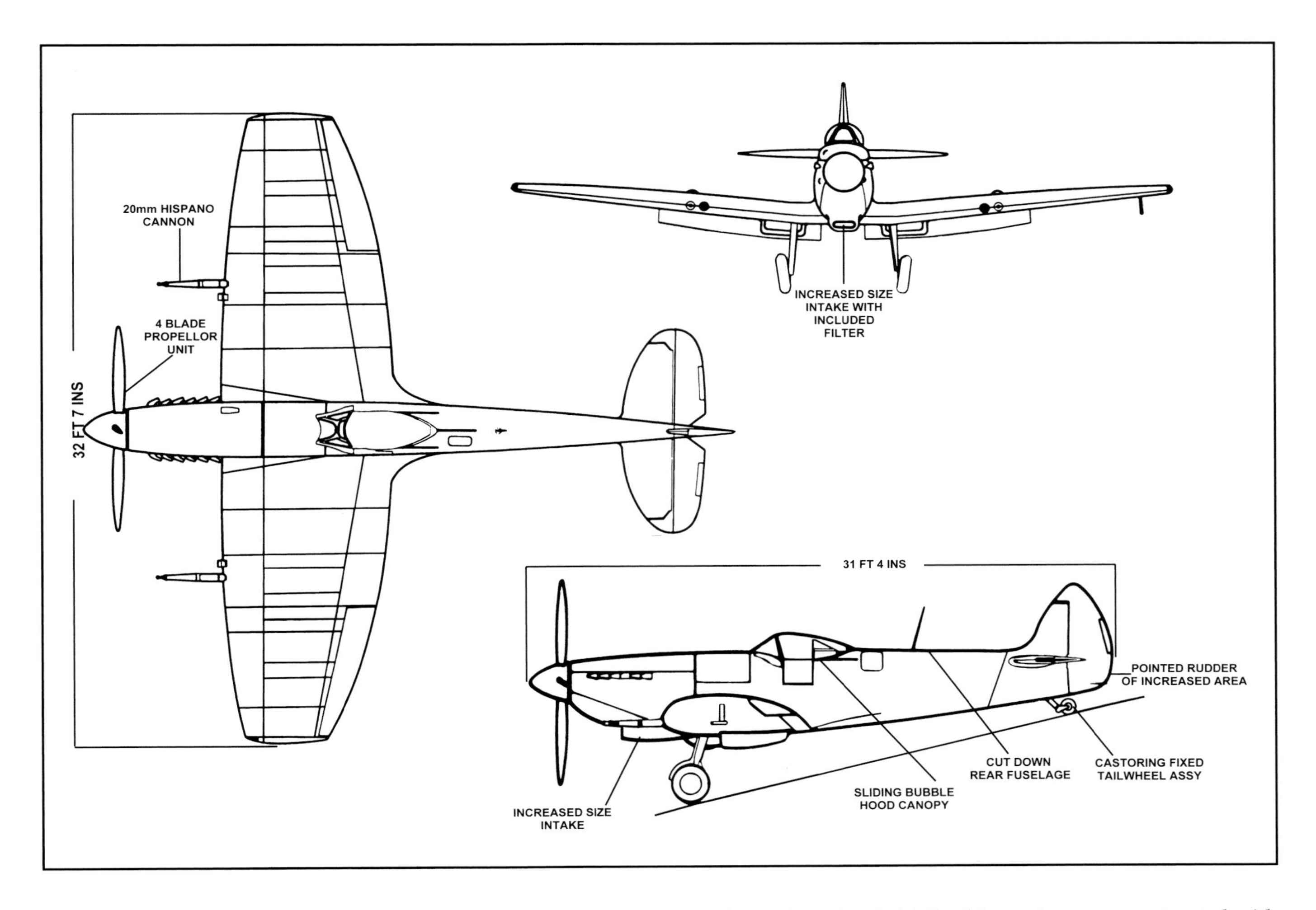

The Spitfire Mk.XVI was essentially a Mk.IX with a Packard-built Merlin 266 engine. Initially this version was constructed with a high-back fuselage although the greater majority had reduced-height rear fuselages and bubble canopies. (BBA Collection)

Combat evaluations of the Spitfire Mk.VIII were undertaken at the AFDU based at Wittering during July 1943. Comparison was made between the test aircraft, JF664, and a standard Mk.IX which was in widespread squadron service. Further development work involved the forthcoming Spitfire Mk.XIV which would see a change of powerplant, the Griffon, and a redefined wing. Supermarine proposed quite a few changes to the wing although only the change in planform and modified leading edge were adopted. Six airframes, serialled JF316 to 321, were allocated to this task, all being fitted with early versions of the Griffon and requiring a large amount of ballast to counter the increased engine weight. Further trials involving these airframes included testing the Rotol contra rotating propeller assembly after which three of the aircraft, JF319 to 321, were fully converted to Mk.XIV standard. Further development work involving the bubble canopy for improved all-round vision for the pilot was tested on Mk.VIII, JF299, and was adopted for general usage on the Spitfire from late 1944. The Spitfire Mk.VIII was also required to act as a fighter bomber. The original installation covered the centreline bomb crutches which were stressed to carry a 500-lbs GP bomb. Additional trials work saw the wing mountings cleared for service although they were limited to a 250-lbs GP bomb each. When all three bomb mounts were occupied, there had to be a reduction in fuel contents to keep the maximum all up weight within certified limits.

Service use of the Spitfire Mk.VIII was extensive in the Middle and Far East although all required that the modified Vokes/Aboukir filter unit be fitted to reduce the ingress of dust and other foreign objects. Designated the Type 360 by Supermarine deliveries to units outside of Europe required that the aircraft were flown by air to their final destination via Gibraltar, Malta, and North Africa. One problem this method of delivery incurred was that each airframe normally arrived

Spitfire ML417 taxies past the camera in its D-Day markings. As the power of the Merlin engine increased, an increase in rudder area was needed thus a more pointed item was fitted. (Sander Wittenaar)

with minimal or no armament, this having to be shipped in by sea. Although a series of overload fuel tanks had been developed to extend the range of the Spitfire, the gains were not enough to allow the carriage of any weapons. Therefore it was not surprising that other methods would be investigated to extend the Spitfire's transit range.

One of the answers postulated was the unique Malianowski wing which was proposed by the Heston Aircraft Company from a design by an engineer of the same name. The idea was to build a fuel tank wing that would be towed behind the fighter thus allowing the aircraft to take off at full weight and armament. Although the top speed would be limited to 250 mph, the available range would be extended to 2,100 miles for each transit leg. Installation was fairly simple as it only required minor strengthening to the aircraft's upper wing surfaces and the installation of universal joints. Fuel transfer was by an electric pump fitted into the overload wing which fed fuel into the Spitfire wing tanks. Attachment to the parent aircraft was by a pair of rigid booms that incorporated explosive charges that could be fired in an emergency.

Delivered to the SAAF after the war this Spitfire LF.IXE, 5581, is complete with Zero length rocket rails which reinforces its low level role. (C P Russell Smith Collection)

First trials were under the aegis of the RAE which commissioned a series of trials from Miles Aircraft using a Magister trainer as the test aircraft. The original wing was built

Complete with a new set of engine cowlings this Spitfire HF.IX poses for its official portrait. It is armed with only a pair of 20mm Hispano cannons. (C P Russell Smith Collection)

with fixed fins and an undercarriage which caused problems when the Magister attempted to take off. In fact, the combination was unable to leave the ground as the wing kept ground looping. Further investigation revealed that articulating the booms and adding controllable rudders fully resolved the ground looping problem and allowed the wing to follow the aircraft. Having solved the problem of the trailing wing, the contract was promptly canceled as it was deduced that underwing tanks would offer a better solution.

The standard Spitfire Mk.IX was normally fitted with a Merlin 61 engine. This aircraft is standing on the engine running pan thus the enlarged tie-barred chocks are required to stop the Spitfire from moving. (C P Russell Smith Collection)

This Spitfire HF.Mk.VIII is a bit of an impostor. It wears the codes of No. 145 Sqdn and the serial MT928, yet it is in fact MV154 which had been built for war service in Australia. Immaculately restored, the aircraft sports a pair of dummy cannons. (Danny Jacquemin)

One of the major users of the Spitfire Mk.VIII was the Royal Australian Air Force of which Nos. 452, 457, and 458 Sqdns used the type with great success against the Japanese opposition throughout the Pacific War. In Europe and the Middle East the Mk.VIII was flown by at least 30 squadrons where they took part in the June 1944 D-Day Invasion and the assault upon Italy. Altogether a total of 1,654 Spitfire Mk.VIIIs were manufactured of all variants. Prior to the cessation of hostilities some examples were passed onto the air forces of Russia and South Africa for evaluation. A further small quantity were employed by the Fleet Air Arm as lead-in trainers prior to the delivery of the Griffon Seafires. After the end of war in Europe and the Far East, quantities of redundant Spitfire Mk.VIIIs were transferred to the Indian Air Force and the French Air Force. Many however never proceeded past the Maintenance Units where they stayed in storage before being finally disposed of between late 1945 and 1947.

The final version of the Spitfire fighter was originally intended as a stopgap before massive deliveries began of the Spitfire Mk.VIII. However, the roles were to be reversed as the Mk.IX was to achieve nearly

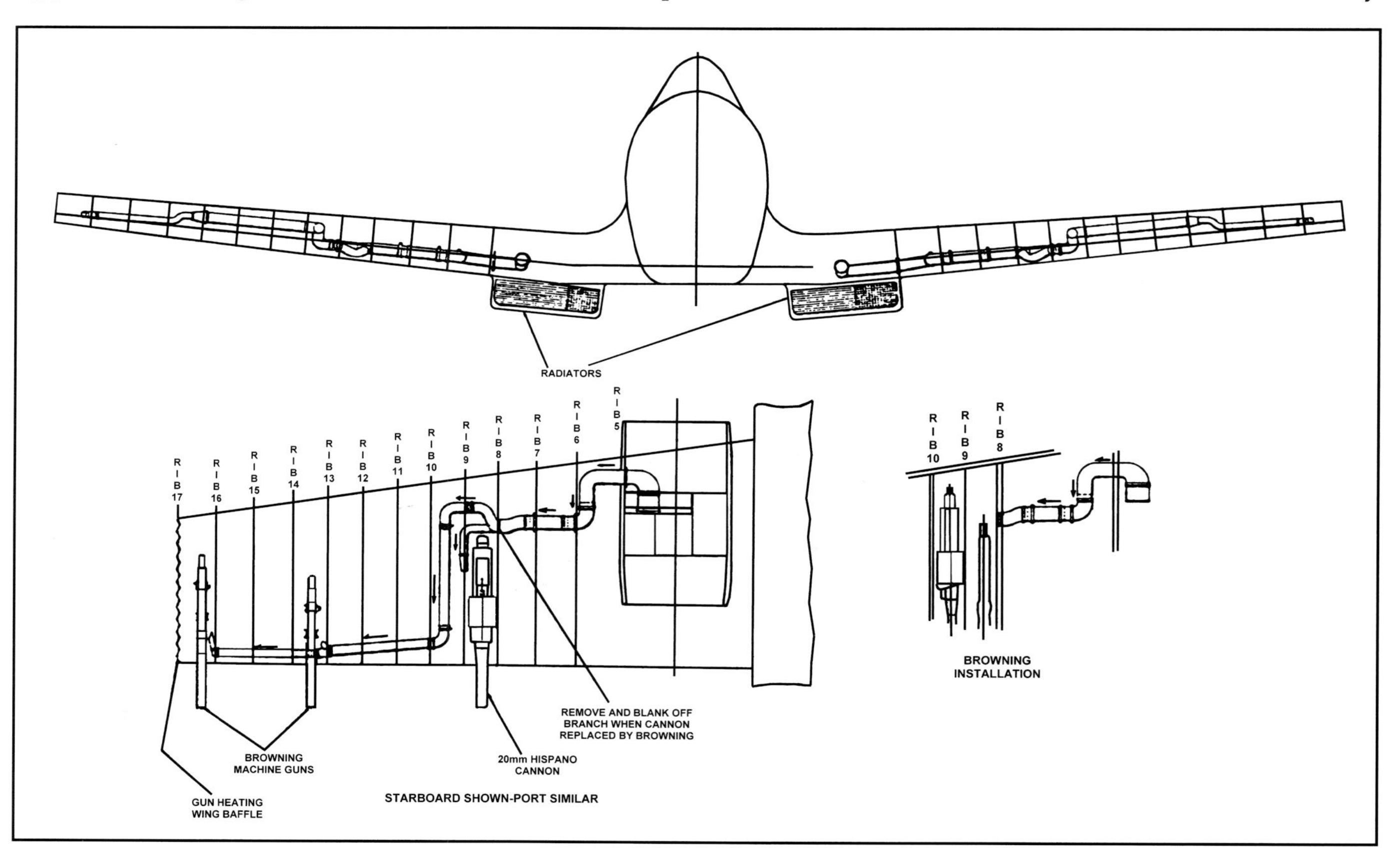

The wing gun heating arrangements could be altered to suit the armament installed as this diagram shows. (via Owen Morris)

11,000 units whilst its Packard Merlin-powered sibling contributed more than another 1,100 separate aircraft. Production of the Spitfire Mk.IX began in early 1942 after a pair of Spitfire Mk.V airframes were sent to Hucknall for fitment of Rolls-Royce Merlin 61 engines. The engine bay bulkhead and primary fuselage longerons were strengthened to accommodate the extra weight, but it was already known by Supermarine and the Air Ministry from the outset that the airframe would require some extra strengthening for combat purposes.

Testing of one of the Spitfire Mk.IX prototypes complete with a Rotol four-blade propeller was undertaken by the AFDU at Duxford during April 1942. All aspects of the report delivered by the evaluators were very positive in nature, including marked improvements in performance over the Spitfire Mk.V at all altitudes. Some months later combat evaluation was undertaken in contest with a captured Focke-Wulf FW-190. Results revealed that both aircraft were fairly evenly matched and that the advantage would be gained by the more experienced pilot.

Another four airframes were quickly diverted to develop the new variant: AA873, with a four-cannon wing; plus AB501, AB505, and AB508. After initial test flying, AB505 was converted to HF.Mk.IX standard and fitted with a Merlin 77 engine, complete with cabin blower which drove a six-bladed contra rotating propeller unit. Air testing began at Hucknall in May 1942 during which period the fin and rudder were modified to the latest standard. All these trials were leading towards the appearance of the HF.Mk.IX which was intended to replace the various versions of the Spitfire originally dedicated to the role. Prior to their appearance a Special Service Flight was formed to operate a handful of stripped aircraft with the task of intercepting the Junkers Ju-86P high flying intruders that had restarted bombing raids on the British mainland.

This Spitfire Mk.IX, MJ175, was on the strength of No. 66 Sqdn when this portrait was taken. Close observation of the aircraft's wheels reveals that it is fitted with nonstandard main wheels. (C P Russell Smith Collection)

Aircraft began rolling from the Supermarine production line in June 1942 with 18 being delivered that month. This was the initial production run of 100 airframes that were converted from Spitfire Mk.V contracts with 52 coming from Supermarine, the rest being a product of Rolls-Royce at Hucknall. Full-blown production of the Spitfire Mk.IX from scratch began immediately after this initial batch had been

A58-484 was a Spitfire HF Mk.VIII operated by the RAAF. Most unusually for an operational fighter it wears a full set of protective covers. Close observation of the panel under the windscreen reveals the victories scored by Group Captain C R Caldwell, the C.O. of No. 80 Wing. (Big Bird Aviation Collection)

Sometimes misidentified as a Mk.XVI this SAAF aircraft, 5621, is in fact a very late-build Spitfire Mk.IXE. (C P Russell Smith Collection)

delivered. The first unit to re-equip was No. 64 Sqdn at Hornchurch being followed by No. 611 Sqdn in July and Nos. 401 and 402 Sqdns in August. In the following month No. 133 (US Eagle) Sqdn received its quota of the new aircraft.

Initially as delivered the Spitfire Mk.IX resembled the Mk.V externally. It still retained the B-type wing, the rounded fin and rudder, and normal sliding canopy. The only obvious visual difference was the change from a semi-circular oil cooler under the port wing to one of a rectangular shape. As the marque developed the wing was changed to that of the universal or E wing, whilst the fin and rudder evolved into a more pointed shape. During the same period the fuselage was cut down and a bubble canopy fitted to improve the pilot's all-round view. This is in turn led to the Spitfire Mk.IXE in both the standard and cut-down fuselage forms capable of carrying either a 250-lbs or a 500-lbs GP bomb. For ferry purposes, a fuel overload tank of up to 90 gallons could be carried instead of the GP bomb.

Although the USAAF had plenty of its own fighters such as the Mustang, they were quite prolific users of the Spitfire. This Mk.IXC, coded EP-A, was the personal mount of Col. E P Allan of the 9th Tactical Air Force based in England. (Big Bird Aviation Collection)

Once Supermarine and Rolls-Royce had delivered the first batch of Spitfire Mk.IXs, the facility at Castle Bromwich began to feed aircraft into the system after retooling from the Mk.V. The first airframe, JK395, appeared on the flightline on 16 March 1943. Other changes introduced by Castle Bromwich included the use of infrared drying lamps in the paint drying shops where the time taken to cure the paint was reduced to minutes instead of hours. Also approved at this time was the IFF Mk.II which was contracted from Ferranti Ltd, but was mainly built in the USA.

Although reasonable quantities of the standard Spitfire Mk.IX and the high-altitude fighter were delivered to the Royal Air Force and its allies, the model produced in the highest quantity was the LF.Mk.IX which was powered by a Merlin 66 with the cabin blower deleted. The first production Spitfire LF.Mk.IX was despatched to Boscombe Down for testing in mid 1943. Most of the test flights concerned bomb delivery as the basic handling qualities of the type had already been quantified. Weapons release could be carried out at all speeds up to maximum although pilots had to be aware of a tendency to pitch up when the bombs were released. Once cleared for production the Spitfire LF. Mk.IX began to leave the various manufacturing facilities in large quantities. Not only did Fighter Command require as many as possible for operations over Europe in the build up towards the Invasion of Normandy, but large quantities were needed for operations in the Mediterranean theatre.

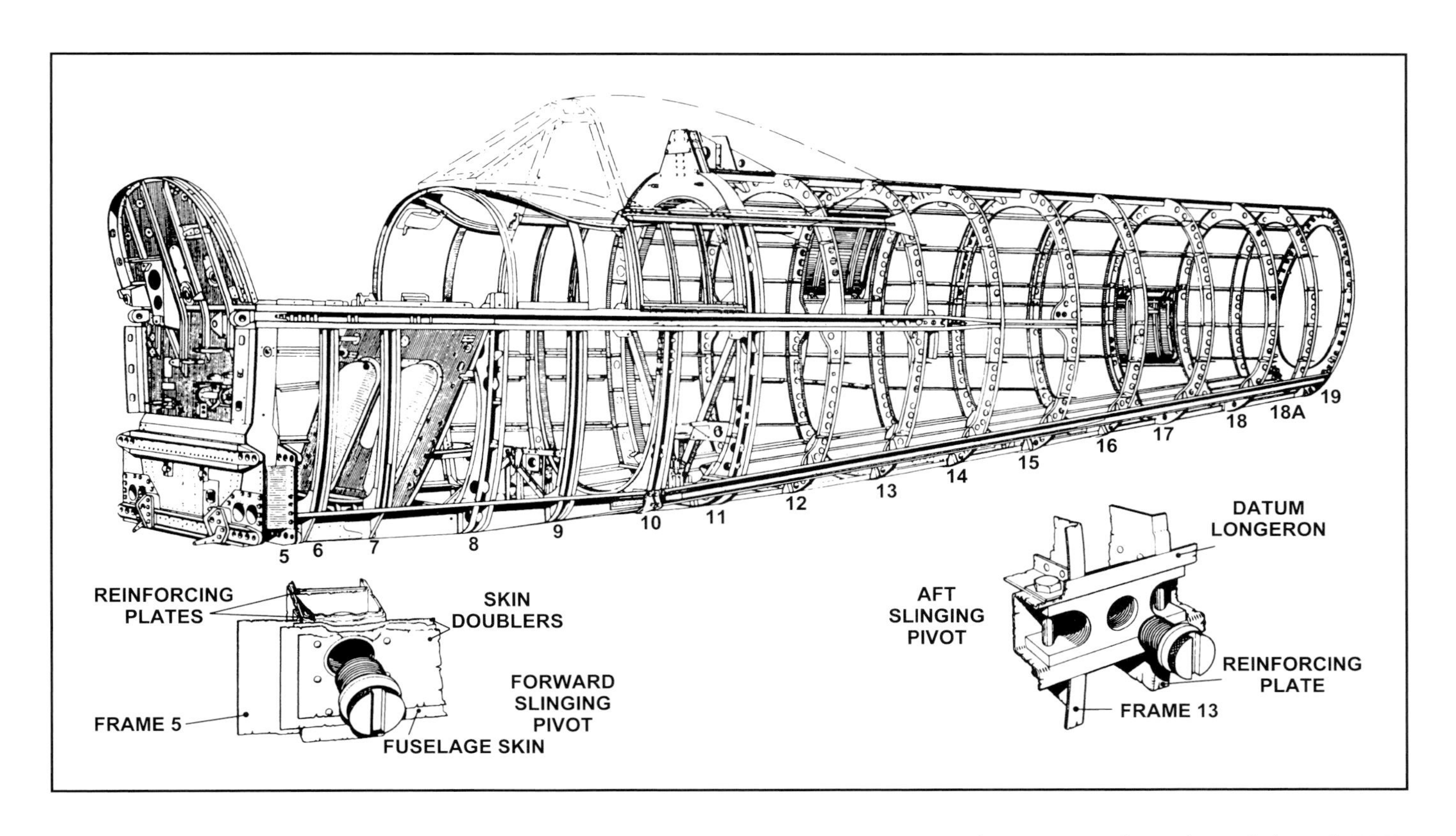

This structural diagram reveals the changes wrought upon the basic Mk.IX airframe to produce that of the Mk.XVI. (Chris Michell)

During operations in both Europe and the Mediterranean, pilots began to notice that the wing upper skins were wrinkling above the cannon and wheel bays. Interviews with numerous pilots elucidated that they were having to manoeuvre their aircraft violently to avoid anti-aircraft defences when attacking vehicle convoys and ground targets. Added to that was the last minute high G forces pull up that was at the

Wearing its wartime codes of 2I is Spitfire Mk.IX, MK356, which underwent a full rebuild at RAF St. Athan prior to joining the Battle of Britain Memorial Flight. (Owen Morris)

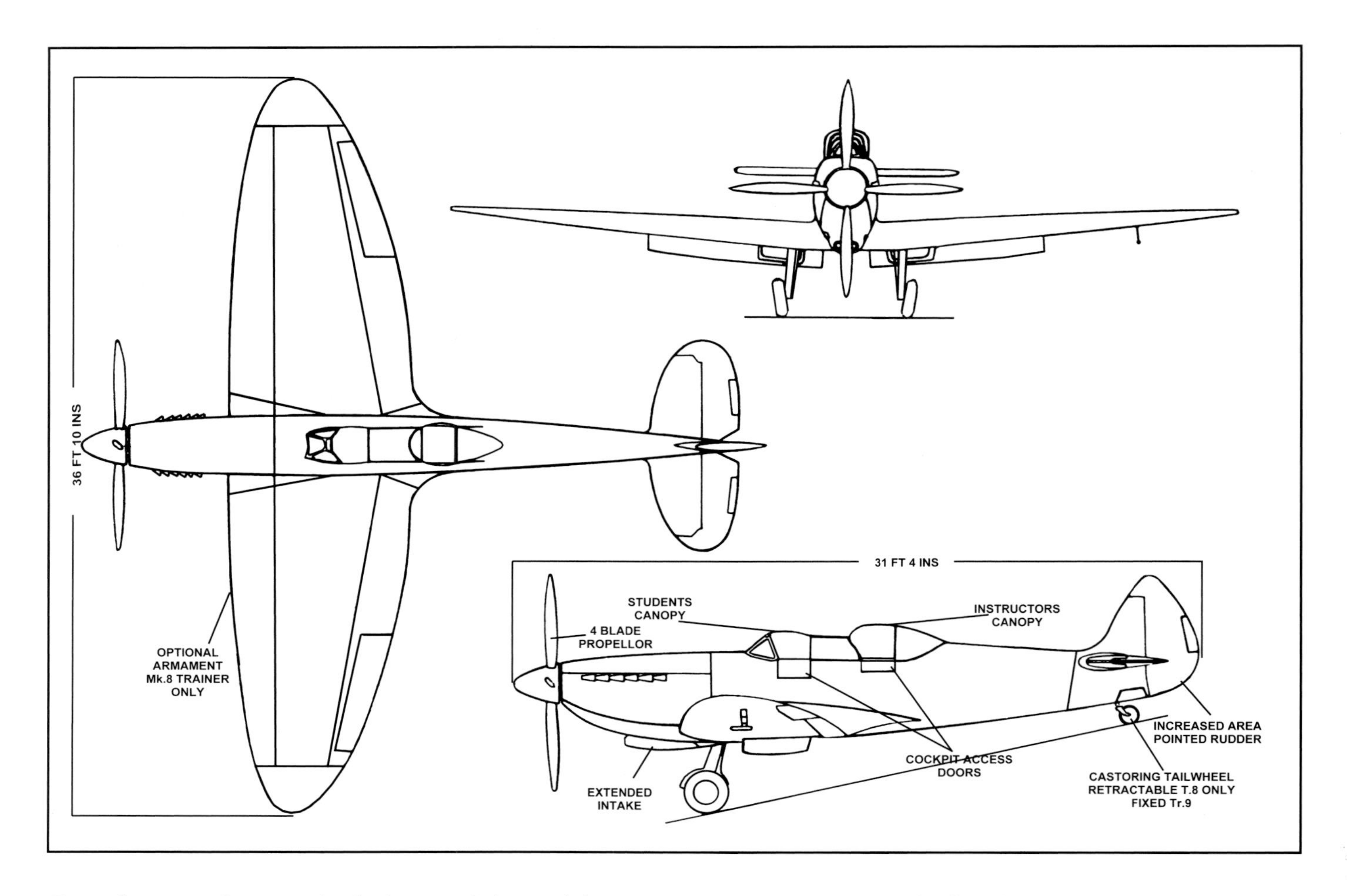

From the outset Supermarine had stressed the need for a two-seat training variant. The first had been a converted Mk.VIII although the majority were reconstructed from redundant late-build Mk.IXs. (Big Bird Aviation Collection)

end of a dive bombing run. Investigation into the structure of the wings of some of these aircraft revealed that the structure remained basically sound and that upper external bracing should be applied if required otherwise the Spitfire was fit for flight.

Converted Spitfire IX, PV202, approaches to land with gear and flaps down. (Big Bird Aviation Collection)

Once the invasion of Europe was underway the role of the Merlin Spitfires in frontline service was almost over. They were being replaced by the Griffon-powered Mk.XIV in the fighter squadrons whilst the reconnaissance squadrons were equipping with the PR.19. This meant that large quantities of Spitfire Mk.IXs were available for use by other air forces. Some of those taking advantage of this surplus included Denmark, Holland, Italy, Israel, Russia, and South Africa. Other aircraft went on to serve postwar with the French Air Force. Still others saw use in Vietnam and took part in the fighting there. Some of the redundant reconnaissance airframes found further service with Egypt and Norway.

With its flaps in the fully deployed position, converted Spitfire Mk.IX touches down after a flight. Unlike "proper" Tr.9s this Spitfire is really only suitable for transport and filming purposes. (Damien Burke)

Very much a derivation of the Spitfire Mk.IX, the Spitfire Mk.XVI was basically the standard airframe with a Packard-built Rolls-Royce Merlin engine designated the Type 266. The reason for the change of designation given to the aircraft enabled spares ordering to be accomplished more accurately. Once initial production difficulties had been overcome the American company eventually delivered 55,000 powerplants with 50% going to Spitfire manufacture at Castle Bromwich. The original Spitfire Mk.XVI as built featured the high-back fuselage inherited from the Mk.IX. However, as technology and the developments in canopy production improved, manufacture changed to that of the cut down version for both the Mk.IX and Mk.XVI.

Performance and handling were very similar to that of its British-powered equivalent and both were regarded as interchangeable. As deliveries to the RAF started late in the war, their operational use was limited and many were delivered straight from the manufacturers to maintenance units for storage. Postwar usage was mainly confined to secondary units within the RAF. A better usage was enjoyed by overseas air forces; the Belgian Air Force received 25 on loan prior to the arrival of Spitfire Mk.XIVs. The only overseas sale was to the Greek Air Force which purchased 54 Spitfire Mk.XVIs.

Spitfire Mk.IX, MK356, eventually ended up at RAF Hawkinge as a gate guard once its flying days were over. Happily it is now restored to flying status with the BoBMF at Coningsby. (C P Russell Smith Collection)

Under less than ideal circumstances Spitfire Mk.IXC, PT934, of No. 331 Sqdn undergoes field maintenance. Raised on jacks for retraction tests on the undercarriage and with cowling panels removed this Spitfire would later resume flying until retired after the war. Close observation to the area aft of the engine reveals the normally hidden top fuel tank. (C P Russell Smith Collection)

The final expression of the Spitfire Mk.VIII and Mk.IX genre was a two-seat trainer. Initially the idea for a two seater was floated when the Mk.I entered service. The Mk.I was sometimes tricky to handle and the Air Ministry believed a two-seat trainer would be needed to speed up the conversion of tyro pilots. The outbreak of war scuppered these plans and they were not to emerge again until the situation had eased and the Mk.VIII emerged.

Prior to Vickers converting Mk.VIII, MT818, in 1946 two others had been produced. One was a local modification of a Mk.V by No. 261 Sqdn in Italy which involved placing an open cockpit in front of the regular one. A more serious effort was carried out in Russia when a proper enclosed

This Spitfire Mk.IX displays all the details that define the late production batches. It sports a tall pointed rudder, standard rounded wingtips and two cannon plus blanking stubs for two others. (C P Russell Smith Collection)

cockpit was built into the fuselage behind the original. Conversion was carried out at Chilbolton, Hampshire on MT818 which first flew in August 1946 registered as N.32. After company flight testing the T.8, now with the registration G-AIDN, was despatched to Boscombe Down for evaluation. The report produced by the test pilots assigned to the aircraft stated that the conversion was easy enough to fly with or without an occupant in the rear seat. However the report was not so kind about the type's use as a conversion trainer. Much was made of the vision problems from the rear seat and the problems that would be encountered should the front canopy be open and the pilot in the rear wished to evacuate the Spitfire. Although strenuous efforts were made to sell the two-seat Spitfire to the RAF as an intermediate between the North American Harvard and the Spitfire of the reserve units, no orders were forthcoming as the intention was to re-equip these units with jet fighters such as the Vampire and the Meteor.

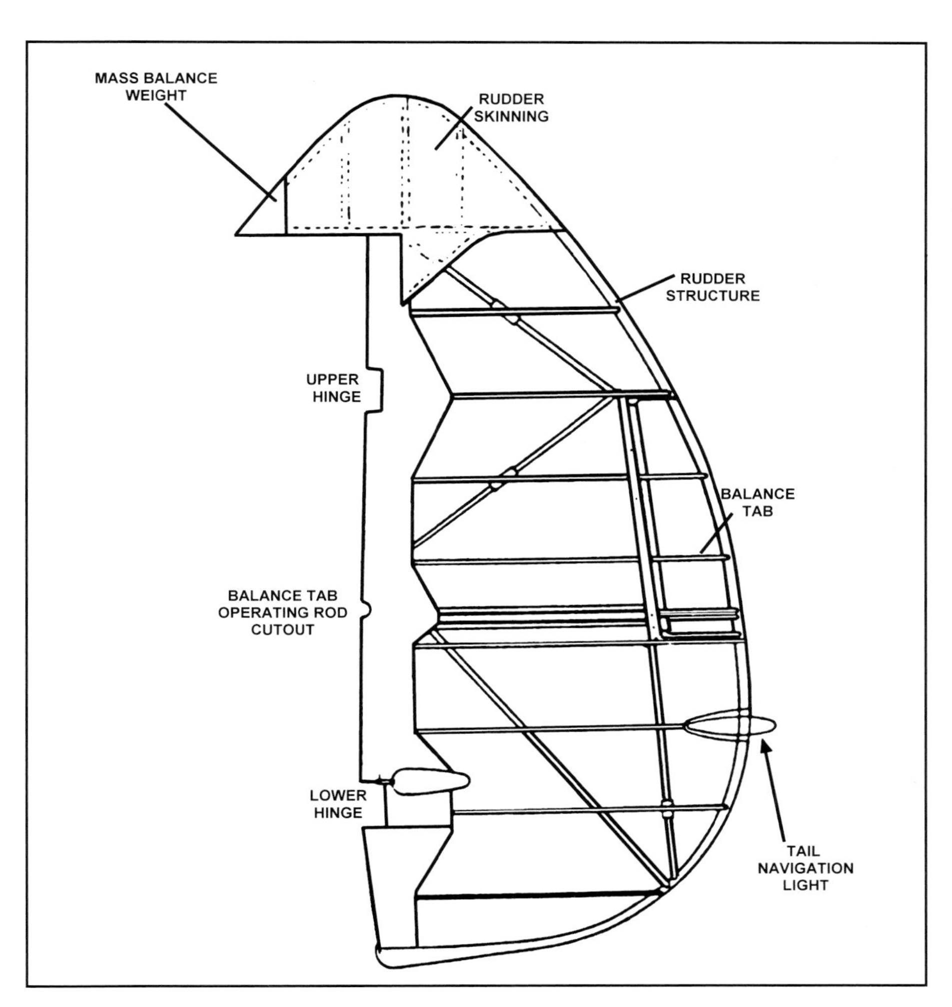

The rudder fitted to the greater majority of Spitfire Mk.VIII, IX, and XVIs was of the broader chord pointed variety. (Chris Michell)

Having failed to interest the RAF in the two-seat trainer Supermarine turned to overseas air forces for sales. Demonstration flights were made by Jeffrey Quill in G-AIDN to numerous air forces some of which expressed an interest in placing orders. Fortunately for Vickers and Supermarine there were vast quantities of unused high-backed Spitfire Mk.IXs and Mk.XVIs in store at the various Maintenance Units throughout the country. Eventually all the hard sales work paid off with contracts being placed by Holland, India, and the Irish Air Forces. Altogether 20 conversions were produced, although the final total could have been higher had the orders from Argentina, Iraq, and Russia been confirmed.

Spitfire ML407 was originally constructed as a standard Mk.IX fighter. After war service it was converted to full Tr.9 standard for use by the Irish Air Corps. When finally demobbed it was sold into private hands, the new owner fitting a more refined rear cockpit. (Big Bird Aviation Collection)

Now resident in the USA is Spitfire Mk.XVIE, TE392, pictured here at No. 5 MU, Kemble, after refurbishment. Complete with clipped wings, bubble canopy, and truncated cannon fairings the aircraft awaits its next duty. (Ray Deacon)

This dismantled Spitfire LF Mk.XVI reveals the details of the wing root assemblies and the propeller hub. (Big Bird Aviation Collection)

Seen from another angle this Spitfire Mk.XVI reveals the clearly defined engine access cowlings. When this photograph was taken the airframe belonged to the RAF Exhibition Flight. (Big Bird Aviation Collection)

Complete with the 8Q codes of No. 695 Sqdn this Spitfire Mk.XVI, TD248, is still extant. As with most postwar Spitfires the cannon armament has been removed. (C P Russell Smith Collection)

Struggling under adverse conditions two members of the ground crew endeavor to service the Merlin engine of a Spitfire Mk.IX. (Big Bird Aviation Collection)

As two engineers struggle to finish fitting the wing spar bolts, all the pipes that festoon a Merlin engine are clearly depicted. (Chris Michell)

With its wings removed this Spitfire Mk.IX sits on jacks and trestles whilst undergoing in-depth maintenance. Removal of the cowling panels has revealed the engine and its ancillary components. (Chris Michell)

Seen on its engine servicing stand is this Merlin currently on display in the Caernarvon Air Museum. (Big Bird Aviation Collection)

Frame 5 is not only the main mounting point for the engine, but also has the wing front spar bolted to it as well. (Chris Michell)

Stripped for repair is the area that includes the port radiator bay. (Chris Michell)

Seen in its assembly stand is this Mk.IX fuselage on its way to completion by Airframe Assemblies on the Isle of Wight. (Chris Michell)

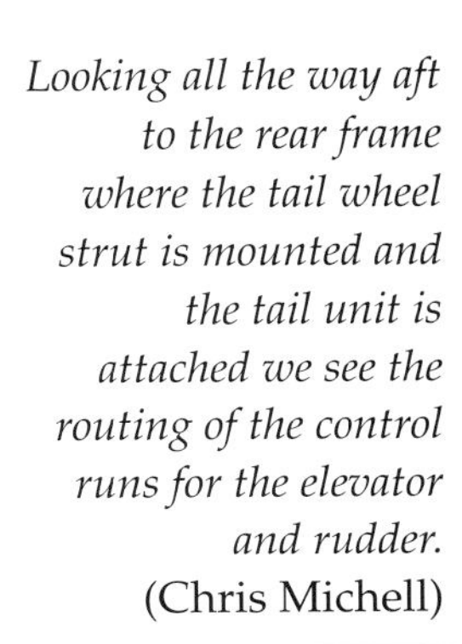

Looking all the way aft to the rear frame where the tail wheel strut is mounted and the tail unit is attached we see the routing of the control runs for the elevator and rudder. (Chris Michell)

This Tr.9 delivered to the IAC was No. 161. Unlike civil-based conversions of the Spitfire currently extant, this variant has a raised cockpit more suited to the instructional role. (C P Russell Smith Collection)

Not only did the Spitfire Mk.XVI serve in the RAF frontline and the R Aux AF, it also became the stalwart of many second line units. TE204 is seen here in the service of No. 3 CAACU based at Exeter. (C P Russell Smith Collection)

Looking worse for wear is TB397, a Mk.XVI allocated to No. 612 Sqdn R Aux AF. Such is the weariness of this airframe that all the panel lines are clearly shown. (C P Russell Smith Collection)

7 Sea Legs

The Merlin Seafires

Although the aircraft carriers of the Royal Navy were fairly modern for the period just prior to the Second World War, the same could not be said of their aircraft complements. The earlier original biplane fighters had been replaced by the Blackburn Roc four-gun turret fighter and Skua strike dive bomber monoplane fighters. However neither of these aircraft would be a match for any of the fighters of the Luftwaffe, given their slow speed and inadequate armament. Even the Fairey Fulmar which was just beginning to enter service with its increased capabilities was seen only as a stopgap.

This was the situation that confronted the Royal Navy and the Fleet Air Arm as they struggled to provide cover for convoys bringing supplies to Britain. Try as they might the Blackburn fighters were unable to provide much opposition to the attacking aircraft of the Luftwaffe. To make matters worse, their attempts to shoot down the shadowing Focke-Wulf FW-200 Condor bomber reconnaissance aircraft frequently ended in failure.

A form of partial relief appeared in the shape of the CAM ship and its single Hawker Hurricane. The downside to this approach was that the Hurricane was a one-use only resource and that the pilot was put in danger as he had to bail out of the aircraft once the sortie was complete. Obviously this waste of men and materials could not be allowed to continue, thus the Admiralty cast about for an answer.

The first aircraft investigated was the Sea Hurricane. This had many advantages in that it had a strong structure that was capable of easy maintenance, a wide track undercarriage that would make deck landing easier, and the ability to carry a wide and useful range of armament. The downside was that the Hurricane was not capable of any increase in speed without major redesign of the flying surfaces. However, Luftwaffe aircraft were improving all the time. There was also the potential threat of an aircraft carrier being built for the German Navy and an accompanying fighter force. For all these reasons the Admiralty was forced to reject the Hurricane.

The only other alternative available was the Supermarine Spitfire which was as comparable as the Hawker aircraft, but was faster and capable of greater development. There was only one area of concern and that was the narrow track undercarriage which could cause problems on a tossing and pitching carrier deck. However, the narrow track was seen as acceptable given the overall capabilities of the Spitfire.

The Admiralty first began its investigations into adapting the Spitfire in November 1939 when an FAA pilot was sanctioned to fly a Spitfire Mk.I. Although the report was favourable no further progress was to be made as Winston Churchill, in his role as Fifth Lord of the Admiralty, vetoed the idea. However delays to new equipment on order from America meant that the urgency for a new fighter had increased.

Thus, in late 1941, a pair of Spitfire Mk.Vs were delivered to the Super-

During 1943 Seafire IIC, MB141, was based at Farnborough for RATOG trials. Always a test airframe it is seen here on take off with the RATOG units installed at the wing roots and operating. (Fleet Air Arm Museum)

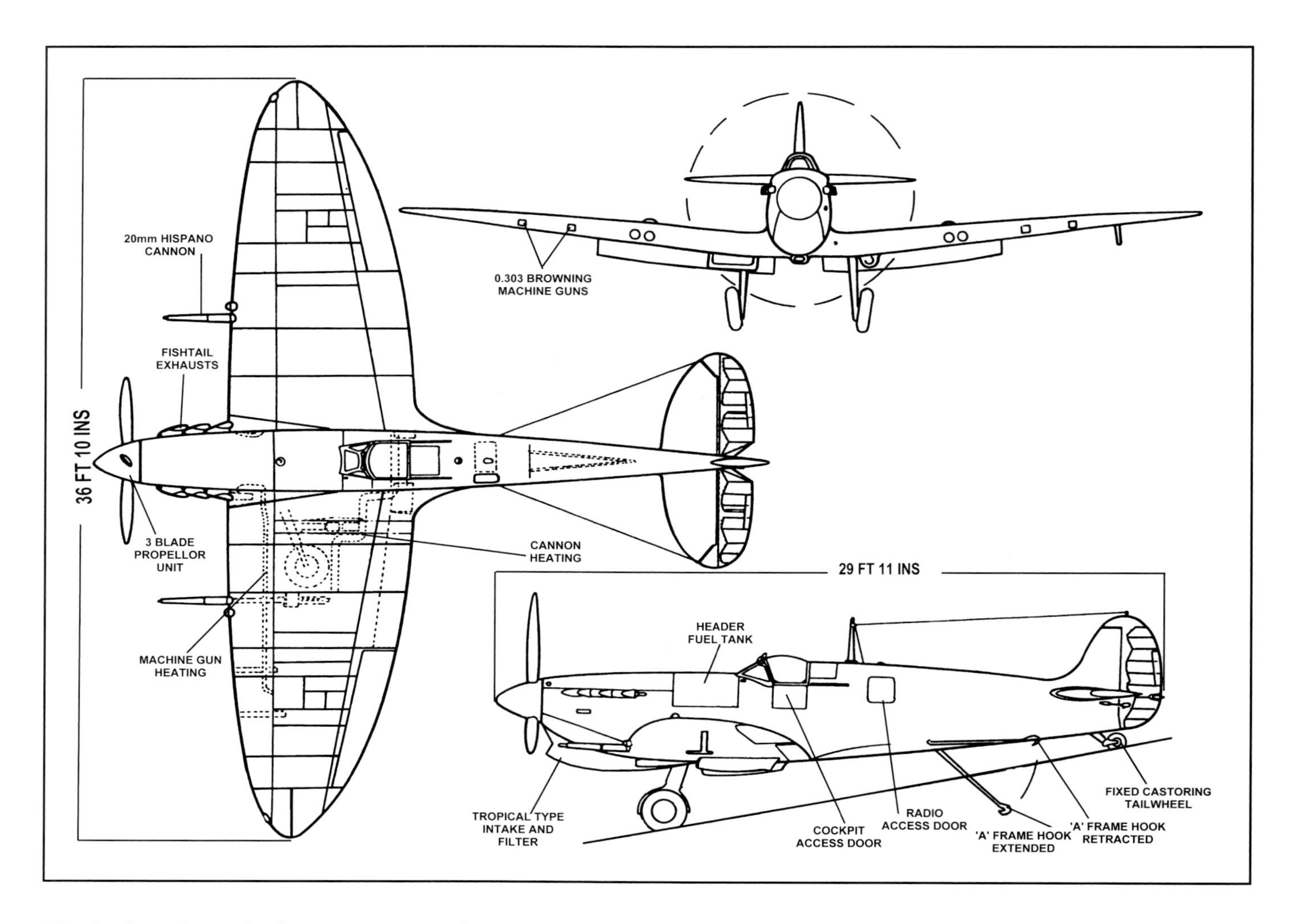

The Seafire Mk.I and Mk.II were very similar in external appearance although the latter had undergone greater navalisation. (Big Bird Aviation Collection)

These Seafire IIC's of No.807 Sqdn are aboard HMS Furious *awaiting orders to take part in Operation Torch. Of note are the open access panels under the wing of the nearest aircraft.* (Fleet Air Arm Museum)

marine airfield at High Post for Navalisation. As these aircraft were intended for training only the modifications embodied into this version, designated the Spitfire Mk.VB (hooked), included an "A" frame arrestor hook and catapult launch spools. The first conversion, AB205, made its maiden flight on 6 January 1942 piloted by Jeffrey Quill. Once Supermarine had completed its flight trials it was despatched to No. 778 Sqdn for naval acceptance trials. Further operational trials were undertaken by the Service Trials Unit which culminated in a landing on the carrier HMS *Indomitable*. Having made this successful landing, a series of trials were undertaken

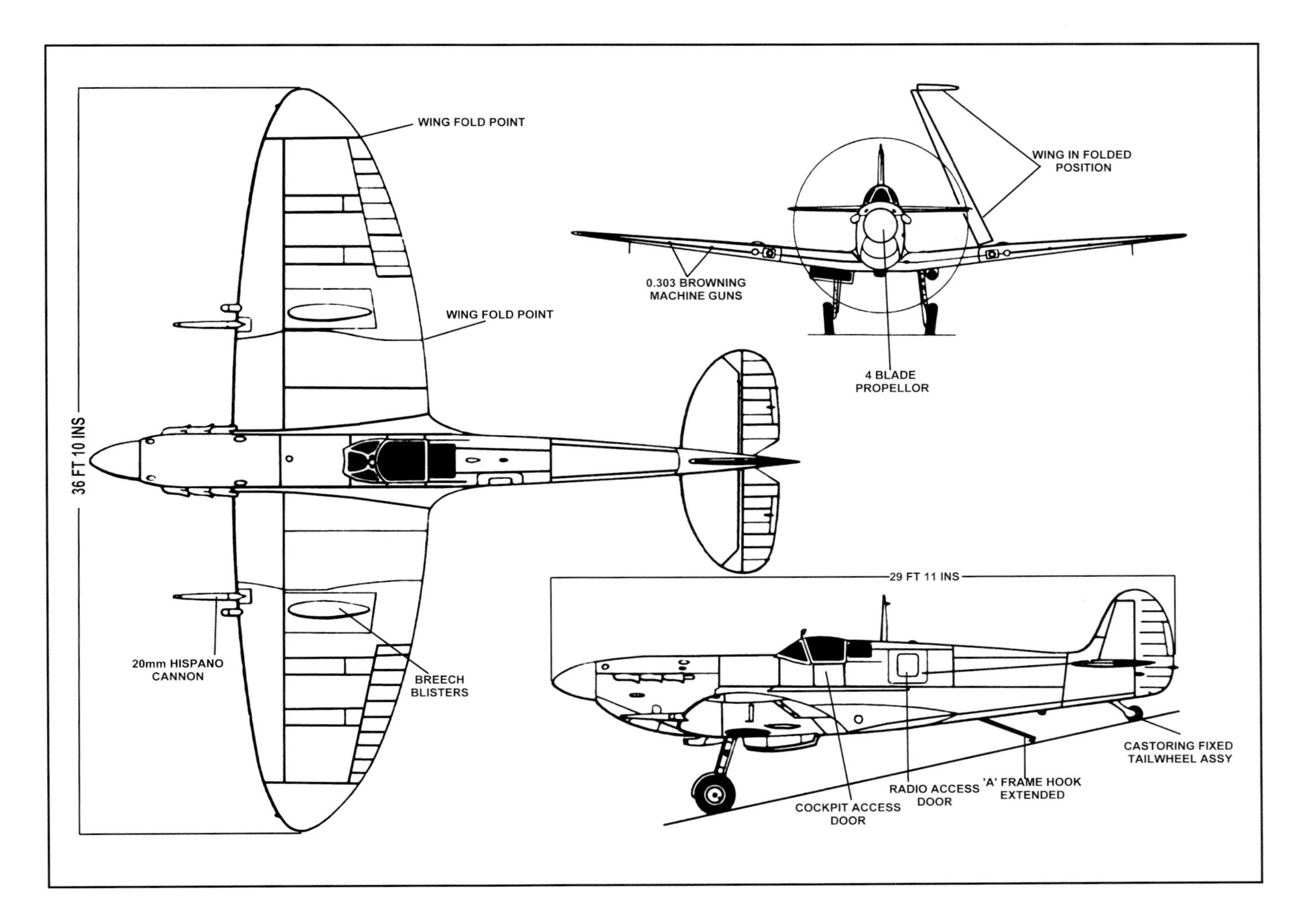

The Seafire F.III was the fully navalised version of the Spitfire complete with folding wings. Modification to this airframe and the addition of a Griffon engine would produce the Seafire Mk.XV. (Big Bird Aviation Collection)

involving assisted take offs, landings, and flybys, all of which came to a stop when the tail wheel mounting sheared upon landing.

This did not, however, stop the Admiralty from placing an order with Air Service Training for the conversion of 48 Spitfire Mk.VBs to Seafire Mk.IB standard which was to feature an arrestor hook only. The next batch of 118 aircraft were ordered from Cunliffe Owen, and not only was an "A" frame arrestor hook installed, but catapult spools were to be fitted also. To counter the stresses placed upon the airframe, the structure needed modifying to improve overall strength. The

Seafire III ,NF490, was allocated to No.728 Sqdn based at Ta Kali, Malta, when this portrait was taken. Due to the dusty climes of that island an extended filter is fitted under the nose. (Fleet Air Arm Museum)

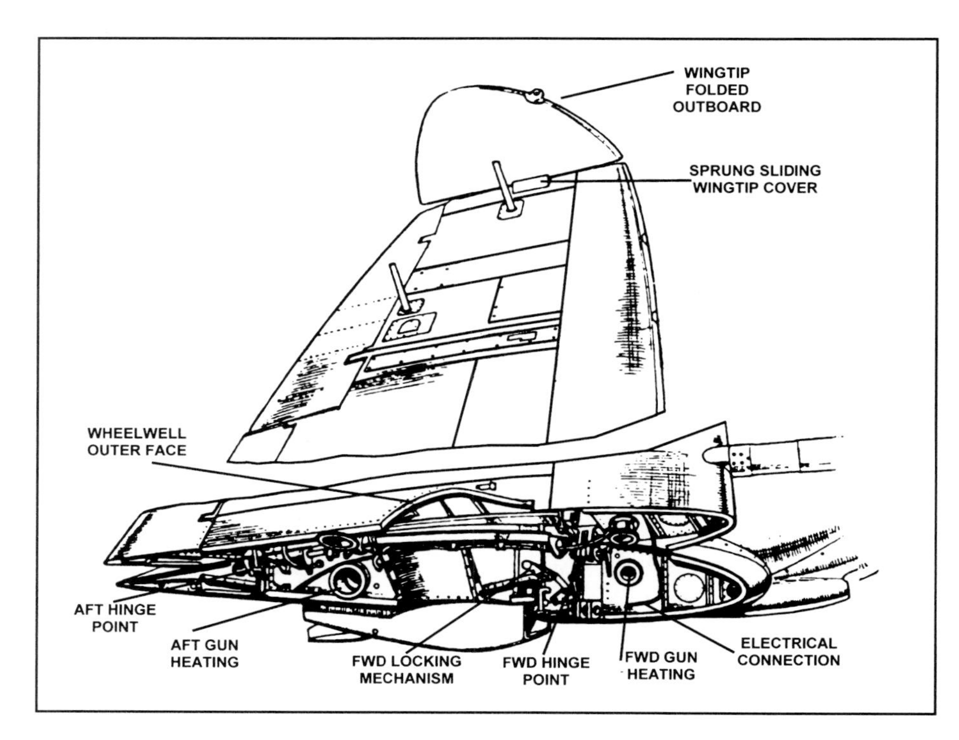

The wing fold arrangement for the Seafire F.III is shown here. A similar system was employed on the early Griffon-powered aircraft. The hinge point is located at the top of the spar whilst locking is achieved through the lower boom. (via FAA Archive)

mounting of the arrestor hook was located at the base of the longerons where they met up with frame 15. This assembly was some six feet long and came with a hydraulic damping system fitted for better control of the hook in the lowered position which extended three feet below the aircraft. In the up position the hook was retained by a scissors-type uplock and was released by pulling upon a Bowden cable release handle. To assist in the loading of the Seafire by crane, reinforced lifting points were installed just behind the engine firewall and behind the canopy.

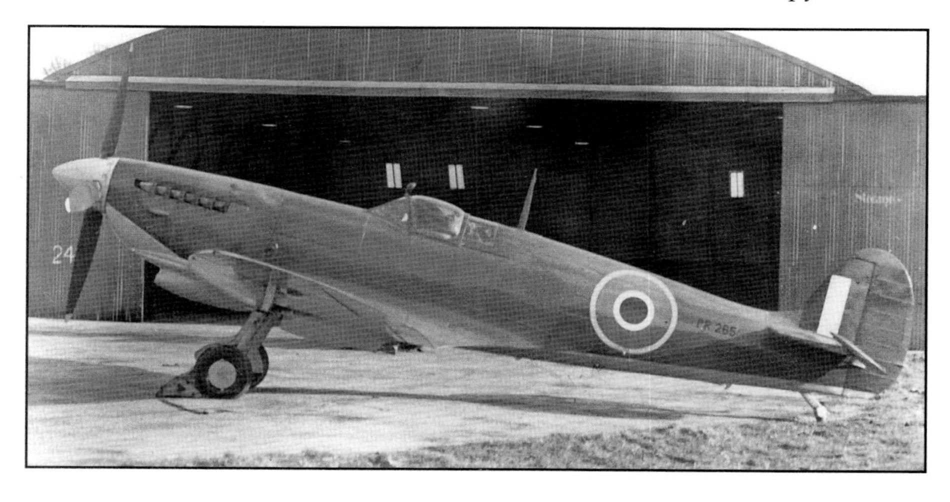

Although this late build Seafire III wears a standard serial number it is in fact awaiting delivery to the French Navy. Close observation of the lower fuselage will reveal the recessed 'A' frame arrestor hook. (Fleet Air Arm Museum)

The first Seafire Mk.IB entered the service of the Fleet Air Arm on 15 June 1942 and undertook carrier trials aboard HMS *Victorious* in the vicinity of the Orkney Islands. These were, as expected, successful therefore further aircraft were cleared for service. Both AST and Cunliffe Owen delivered their completed aircraft via Nos. 38 and 76 MU which carried out final preparations for shipboard use. The primary user of the Seafire IB was No. 801 Sqdn embarked on HMS *Furious* in October 1942 where they remained until October 1944. Unlike most of her contemporaries HMS *Furious* was equipped with T-shaped lifts and large hangars capable of housing the Seafire which had no wing fold. No. 842 Sqdn also received some Seafire Mk.IBs in mid 1943 and operated them from HMS *Fencer.* Other units also flew the type in small quantities, including Nos. 1 and 2 Naval Fighter Schools based at Henstridge and Yeovilton respectively, as did the School of Naval Warfare based at St. Merryn. Examples were also on the strength of the Naval air stations at Lee-on-Solent, Stretton and at Yeoviltons No. 760 Sqdn.

Some of the Seafire Mk.IBs were used for trials and evaluation purposes. One of the first being MB328 which underwent handling and fuel system trials, followed by trials of a 30-gallon centreline fuel tank. Following on from this the aircraft underwent a series of evaluation flights covering the behaviour of an F.24 camera installation and the Hispano 20mm cannon installation.

The Seafire Mk.IB left frontline service in late 1943 with most examples going on to fulfill a variety of training roles. At least one was flown from HMS *Pretoria Castle* for turbulence trials in an effort to avoid pilots

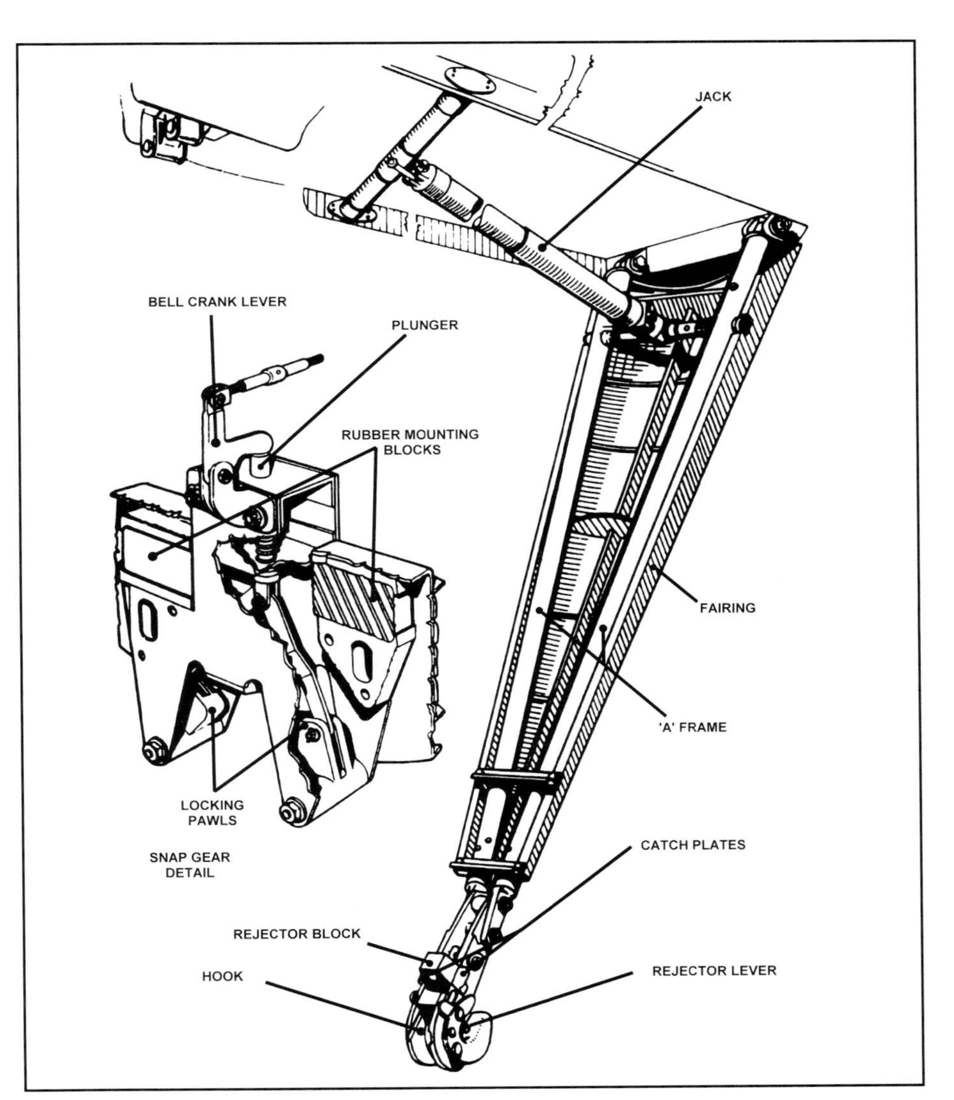

The primary method of retardation aboard an aircraft carrier is the arrestor hook. This is the A-frame type as fitted to the early Seafires. Initially it was rated at 7,000 lbs strain which was later increased to 10,000 lbs. (Big Bird Aviation Collection)

flying into the dead zone aft of an aircraft carrier and stalling into the sea due to the sudden lack of lift. As well as the production Seafires there were batches of Spitfires delivered to the FAA which included 44 Mk.VBs and 55 Spitfire Mk.VBs (hooked).

Following on from the interim Seafire Mk.IB the Admiralty requested that a more advanced version equivalent to the Spitfire Mk.V be prepared for use by the Fleet Air Arm. The forthcoming long-range Fleet fighter, the Firefly, was experiencing development delays and would therefore be late entering service. As the Air Ministry was in charge of Spitfire and the subsidiary Seafire production, it offered some stored Spitfire Mk.Is and IIs as a short-term measure as all available Spitfire Mk.Vs were needed to reinforce the RAF and provide much needed fighters for the Soviet Union. In reply the Navy insisted that it needed at least 430 examples with an attrition follow up of 20 aircraft per month. A review of the Fleet Air Arms requirements was performed and the report referred to the delays being experienced by the future fighters under design and development. Eventually common sense prevailed and the production of the Seafire Mk.IIB was authorised.

Although Supermarine had already carried out the majority of the redesign work required to produce the next Seafire variant, the production contract was issued to Westland Aircraft based at Yeovil. This course

This unidentified Seafire IIC is about to touch down on HMS Furious. *The earlier 'A' frame hook is ready to catch the wire and the flaps, used only for landing, are fully deployed.* (Fleet Air Arm Museum)

of action would in turn leave Supermarine free to concentrate upon the development of the Spitfire which was seen as its primary task. Design changes from the earlier Seafire Mk.IB were quite extensive and included strengthening for the arrestor hook and changes to its housing. The catapult spools and their mounts also came in for serious redesign as increased strengthening was applied to the longerons about frame 16. Further internal structural strengthening was applied to frames 9, 10, and 16 as well as the main fuselage longerons. Increased armour plate also featured in the Seafire Mk.II as did increased strengthening to the Universal C-type wings which allowed an increase in the amount of external stores that could be carried. Modifications were also undertaken to the main undercarriage units which were raked forward slightly in order to reduce the incidences of nose over landings which the Seafire Mk.IBs were prone to.

Designated by Supermarine as the Type 357, the first Seafire Mk.IIC was in fact a converted Mk.V, AD371, which made its maiden flight in February 1942. After initial company shakedown flights the Seafire was transferred to Farnborough for evaluation. During the months that followed AD371 underwent extensive carrier trials aboard HMS *Illustrious* when some faults were discovered in the aircraft's structure. Most of the faults were traced to the stresses generated upon the catapult spools during launch. Frame 19 was the main zone to suffer especially on the starboard side of the fuselage, and the tailplane was found to have a definite twist to starboard. Rectification involved reinforcing the tailplane spar plus other areas of the fuselage.

With these modifications incorporated Supermarine was able to roll out the first production example, MA970, in May 1942, complete with the B wing armament of two cannon and four Browning machine guns. This particular setup had been forced upon the Admiralty as the preferred option since four cannon was deemed too heavy for carrier usage. Although Supermarine geared up quickly to build the Seafire II, the other production line at Westland Aircraft took longer to setup and the first production airframe took until December 1943 to appear.

Captured entering Singapore harbour in September 1945 is the aircraft carrier HMS Hunter *whose fighter unit was No.807 Sqdn. The version of Seafire used by this unit was the LF IIIC. The nearest aircraft is NN300 coded D-5O which is securely lashed to the deck.* (Fleet Air Arm Museum)

As delivered the first 50 airframes were powered by the Merlin 45 engine which reached its limitations at 13,000 feet. Fortunately for the Fleet Air Arm most combat engagements were taking place at lower altitudes, although more power in this part of the flight envelope would be appreciated. This was forthcoming in the form of the Merlin 32 which increased the aircraft's overall performance at lower levels. Adding a four-bladed propeller and a cropped supercharger impeller created an aircraft fully suited to its task. The resultant modifications meant a redesignation, in this case to L IIC. One final modification that had a positive effect on the Seafire was the change of starting method which now utilised the Coffman starter cartridge system instead of the electrical accumulator trolley or mains power units.

In common with their land-based counterparts, some of the Seafires had their wingtips clipped becoming LF.IICs in the process. This had the same result as that of the Spitfire, a faster rate of roll although maximum altitude was restricted to 20,000 feet. This reduction in span also had other effects, none of which were particularly beneficial. These included an extended take off run, a higher landing speed and a need for an extended runway. Allied to this was the difficulty in handling when the aircraft was fully laden. Overall the advantages gained were beneficial although the accident rate with this

The aircraft in this formation are all Seafire IB's which was the next stage of development after the hooked Spitfire. The nearest Seafire is NX890 and all three come from No.736 Sqdn. (Fleet Air Arm Museum)

variant accounted for a third of total Seafire crashes. Attempts were made to provide some improved control during the landing phase. A tail braking chute was introduced which performed reasonably well, but was abandoned due to the technical difficulties involved in its installation. The next idea tested covered alterations to the flap system to provide the Seafire with a take off flap setting, the norm being that flaps were used for landing only. This was yet another modification that was eventually abandoned due to the technical difficulties involved.

The final version of the Seafire II developed for the FAA was the reconnaissance fitted LR.IIC which touted a pair of F.24 cameras. One camera had a focal lens length of 20 inches and was mounted in the vertical plane whilst its oblique counterpart came equipped with a lens of 14 inches in focal length. The LR.IIC retained its full armament options as it was regarded as a combat aircraft. Conversion was carried out by Heston Aircraft and the first modified aircraft made its maiden flight in July 1943. The final Seafire II was delivered in October 1943 after 787 examples had been manufactured.

In Fleet Air Arm service the Seafire II in its various versions served with No. 807 Sqdn which completed its conversion fully to the type in May 1943. Other units to be fully or partially equipped with the marque included Nos. 807, 809, 834, 879, 880, 884, and 894 Sqdns. This size of force was required in light of the task that faced the Royal Navy during the invasion of Italy. Code named Operation Avalanche the assault upon Italy began on 9 September 1943 and included the Aircraft Carriers HMS *Attacker*, HMS *Battler*, HMS *Hunter*, HMS *Stalker*, and HMS *Unicorn*. On board these vessels were a total of 106 aircraft. By the time operations ceased in support of the landings, over half of the available aircraft had been lost due to enemy action and landing mishaps. Once their front-line career was over many of the survivors were cascaded to second line units such as No. 761 Sqdn where they were used in the training role.

The final version of the Merlin-powered Seafire to be developed was the aircraft that the Admiralty had been after since the beginning. Designated the Mk.III a total of 1,218 would be procured as well as some hybrid Mk.IICs of which 32 were converted.

Initial work had involved literally marking lines in chalk over the wings of a Spitfire Mk.I on loan to General Aircraft Ltd in the summer 1942. Working in conjunction with a set of structural drawings the fold points were defined as just outboard of the main undercarriage bay and at the wingtip fitment break point. The results garnered were translated into metal using the first production Seafire Mk.II, MA970, as the prototype conversion. Work began on adapting the wings in October 1942 and the aircraft was ready for flight trials in early November.

The inner primary hinging point was located on the top of each boom of the main spar whilst the lower boom housed the locking pin. On deployment of the wings to the locked position the openings in the upper skin surface were covered by spring-loaded fairing panels, the whole being positively locked by an operating lever in each wheel bay. To guide the wing into position there were spigots on the movable section that located into sockets in the fixed portion whilst the wing heating trunking interfaced using felt jointing pads. In this first incarnation the wing fold was purely manual therefore the space saving of the wing fold mechanism was offset by the requirement for extra personnel to fold and deploy the mainplanes.

The extra weight of the wing fold mechanism meant that weight saving had to be achieved in some other area. This meant that the outer cannon gun bays had to be deleted. There was an advantage in this in that the blisters covering the breech and feed mechanisms could be made smaller thus increasing the speed by a few miles per hour. As a retrospect some already built Seafire Mk.IICs had this modification embodied. To drive the new version of the Seafire, the Merlin 55 engine was selected. This version was an adaptation of the Merlin 45 with the same output power but featured automatic boost control complete with barometric governor. On the main production batch this drove a four-bladed Rotol propeller although a few of the early build were rolled out with three-bladed units.

The prime contractor for the production of the Seafire Mk.III was Westland Aircraft based at Yeovil with Cunliffe Owen Ltd acting as the primary sub contractor. Difficulties were initially experienced in starting production thus the first few aircraft were delivered as Seafire F.IICs. Flight trials of the type were carried out during June 1943. The type's performance was assessed as satisfactory although a mention about the deterioration in the paint finish was made and passed to all those involved in production. Further problems arose as the flight trials progressed when a discrepancy arose in the aft C of G datum point and caused some instability in the air. In-depth investigation revealed that the original suspect, the tail unit, was in fact having its airflow disturbed by the wing fold joint.

Soon after entering service pilots began to report that there seemed to be a problem with the gun bay heating, therefore remedial action took the form of asbestos blanketing and the installation of a baffle plate. Troubles were also being experienced during the landing phase as the arrestor hook appeared to bounce back into the airframe. The first stage of remedial action involved strengthening the surrounding structure which had some effect although the problem was still regularly encountered. Therefore, a series of trials with a selection of dampers was carried out. Allied to the hook bounce was the discovery of tip damage to the propeller assemblies. Careful study of Seafire Mk.IIIs landings showed that there was an oscillating bounce on touch down which caused the nose to pitch down. Improvements to the undercarriage shock absorber helped reduce this tendency although further modification to the down locking mechanisms was required to counter a tendency for the undercarriage legs to either collapse or distort upon landing at an offset angle.

The Seafire Mk.III was also subject to development trials to extend its capabilities, thus the wings were strengthened to allow the carriage of rockets and bombs up to a maximum of 500 lbs. Improvements were also undertaken to the primary cannon armament where the original Hispano Mk.IIs were replaced by the shorter and therefore lighter Mk.Vs. Unsurprisingly the Seafire Mk.III also appeared in the reconnaissance format and was equipped with a pair of F.24 cameras, one set in the vertical plane whilst the other was set at an oblique angle. Eventually a total of 129 of this versatile marque were constructed. Altogether a final total of 1,220 of the Seafire III were built and equipped most of the squadrons of the Fleet Air Arm, both at home and in the Far East.

Foreign sales included 12 to the Irish Air Corps, although these were fully denavalised before delivery, and 48 to the French Navy. The Irish aircraft survived in service until the mid 1950s until they were replaced by the piston Provost.

Appendix A: Spitfire Data

Important Characteristics

Marque	Status	Powerplant	Propeller	Armament	Span	Length	Remarks
Proto	Prototype	Merlin C	2 blade	Nil	36ft 10ins	29ft 11ins	Tail skid, Cream finish
Proto	Modified	Merlin F	2 blade	Nil	36ft 10ins	29ft 11ins	Re-engined for trials
Proto	Modified	Merlin II	2 blade	A	36ft 10ins	29ft 11ins	Modified to Mk.I
Mk.I	Early Prod	Merlin II	2 blade	A	36ft 10ins	29ft 11ins	1st production
Mk.I	Production	Merlin II	3 blade	A	36ft 10ins	29ft 11ins	Modified production
Mk.I	Production	Merlin III	3 blade	A	36ft 10ins	29ft 11ins	Modified production
Mk.I	Experimental	Merlin RM2S	3 blade	deleted	36ft 10 ins	29ft 11ins	Engine test K9788 by RR
Mk.I	Experimental	Merlin RM3S	3 blade	deleted	36ft 10 ins	29ft 11ins	Engine test K6889 by RR
Mk.I	Experimental	Merlin RM4S	3 blade	deleted	36ft 10ins	29ft 11ins	Engine tests K9877 and
Mk.I	Experimental	Merlin RM5S	3 blade	deleted	36ft 10ins	29ft 11ins	N3053 used by RR
Mk.I	Export	Merlin III	3 blade	A	36ft 10ins	29ft 11ins	Greece export
Mk.I	Export	Merlin III	3 blade	A	36ft 10ins	29ft 11ins	Portugal export
Mk.I	Export	Merlin III	3 blade	not fitted	36ft 10ins	29ft 11ins	Turkey export, 2 delivered
Mk.I	Modified	Merlin III	3 blade	A	32ft 2ins	29ft 11ins	Clipped wing trials
Mk.I A	Modified	Merlin III	3 blade	A	36ft 10ins	29ft 11ins	40-gal L.R. Tank port wing
Mk.IB	Modified	Merlin III	3 blade	2x Cannon	36ft 10ins	29ft 11ins	Cannon aircraft
Mk.I	Modified	Merlin III	3 blade	A	36ft 10ins	29ft 11ins	Float plane, Roc floats
Mk.I	Modified	Merlin IIIm	4 blade	deleted	33 ft 8 ins	29ft 11ins	Speed Spitfire K9834/N.17
Type A	RAF Mod	Merlin III	3 blade	deleted	36ft 10ins	29ft 11ins	PR aircraft, 2 only
Type B	PR conv	Merlin III	3 blade	deleted	36ft 10ins	29ft 11ins	PR aircraft with 29-gal tank
Type C	PR conv	Merlin III	3 blade	deleted	36ft 10 ins	29ft 11ins	PR development
Type D	PR conv	Merlin II	3 blade	deleted	36ft 10ins	29ft 11ins	Development of C, 2 only
Mk.IIA	Conversion	Merlin XII	3 blade	A	36ft 10ins	29ft 11ins	Conversion from Mk.I
Mk.IIA	Production	Merlin XII	3 blade	A	36ft 10ins	29ft 11ins	750 built
Mk.IIA	Modified	Merlin XII	3 blade	A	32ft 2ins	29ft 11ins	Clipped wings few built
Mk.IIA	Experimental	Merlin XII	3 blade	A	36ft 10ins	29ft 11ins	Flush fitted oil/fuel tanks
Mk.IIA	Conversion	Merlin XII	3 blade	A	36ft 10ins	29ft 11ins	Conversion from Mk.I
Mk.IIA	Modified	Merlin XII	3 blade	deleted	36ft 10ins	29ft 11in	Mod for Met photo P7301
Mk.IIA	Conversion	Merlin XII	3 blade	A	36ft 10ins	29ft 11ins	Conversion from Mk.I
Mk.IIA	Conversion	Merlin XII	3 blade	deleted	36ft 10ins	29ft 11ins	Mod for Met recording
Mk.IIA	Conversion	Merlin 47	4 blade	A	36ft 10ins	29ft 11ins	Engine testbed HF.VI
Mk.IIB	Production	Merlin XII	3 blade	B	36ft 10ins	29ft 11ins	170 built
Mk.IIB	Modified	Merlin XII	3 blade	deleted	36ft 10ins	29ft 11ins	Experimental
Mk.IIC	Modified	Merlin XII	3 blade	A	36ft 10ins	29ft 11ins	IIA for ASR duties
Mk.IIB	Modified	Merlin XII	3 blade	A	36ft 10ins	29ft 11ins	Clipped wings, 1 only
PR III	Redesignation of Mk.I(PR) aircraft						
Mk.III	Experimental	Merlin XX	3 blade	none	30ft 6ins	29ft 11ins	Clipped wings N3297
Mk.III	Experimental	Merlin XX	3 blade	none	36ft 10ins	29ft 11ins	Standard wings N3297
Mk.III	Experimental	Merlin 60	4 blade	none	33ft 9ins	29ft 11ins	Modified wing tips
Mk.III	Experimental	Merlin 61	4 blade	none	30ft 6ins	29ft 11ins	Trial engine installations
Mk.III	Experimental	Merlin XX	3 blade	various	32ft 2ins	29ft 11ins	Armament trials
Mk.III	Prod project	Merlin XX	3 blade	B	30ft 6ins	29ft 11ins	750 ordered, became Mk.V

Marque	Status	Powerplant	Propeller	Armament	Span	Length	Remarks
PR.IV	Production	Merlin 45	3 blade	none	36ft 10ins	29ft 11ins	229 built in total
PR.IV	Production	Merlin 46	3 blade	none	36ft 10ins	29ft 11ins	As above
PR.IV	Production	Merlin 50	3 blade	none	36ft 10ins	29ft 11ins	As above
PR.IV	Production	Merlin 50A	3 blade	none	36ft 10ins	29ft 11ins	As above
PR.IV	Production	Merlin 55	3 blade	none	36ft 10ins	29ft 11ins	As above
PR.IV	Production	Merlin 56	3 blade	none	36ft 10ins	29ft 11ins	As above
Mk.VA	Conversion	Merlin 45	3 blade	A	36ft 10ins	29ft 11ins	Conv from MkI and II
Mk.VA	Production	Merlin 45	3 blade	A	36ft 10ins	29ft 11ins	Developed from Mk.I/II
Mk.VA	Production	Merlin 45M	3 blade	A	32ft 2ins	29ft 11ins	Delivered as LF.VA
Mk.VA	Production	Merlin 46	3 blade	A	36ft 10ins	29ft 11ins	Developed from Mk.I/II
Mk.VA	Production	Merlin 50A	3 blade	A	36ft 10ins	29ft 11ins	Developed from Mk.I/II
Mk.VA	Production	Merlin 50M	3 blade	A	32ft 2ins	29ft 11ins	Delivered as LF.VA
Mk.VA	Production	Merlin 56	3 blade	A	36ft 10ins	29ft 11ins	Developed from Mk.I/II
Mk.VA	Production	Merlin 45	3 blade	removed	36ft 10ins	29ft 11ins	glider tow trials
Mk.VA	Production	Merlin 45M	3 blade	A	32ft 2ins	29ft 11ins	Delivered as LF VA
Mk.VA	Production	Merlin 50M	3 blade	A	32ft 2ins	29ft 11ins	Delivered as LF VA
Mk.VA	Production	Merlin 55M	3 blade	A	32ft 2ins	29ft 11ins	Delivered as LF VA
Mk.VB	Conversion	Merlin 45	3 blade	B	36ft 10ins	29ft 11ins	Developed from Mk.I/II
Mk.VB	Production	Merlin 45	3 blade	B	36ft 10ins	29ft 11ins	Developed from Mk.I/II
Mk.VB	Production	Merlin 45M	3 blade	B	32ft 2ins	29ft 11ins	Delivered as LF VB
Mk.VB	Production	Merlin 46	3 blade	B	36ft 10ins	29ft 11ins	Developed from Mk.I/II
Mk.VB	Production	Merlin 50	3 blade	B	36ft 10ins	29ft 11ins	Developed from Mk.I/II
Mk.VB	Production	Merlin 50A	3 blade	B	36ft 10ins	29ft 11ins	Developed from Mk.I/II
Mk.VB	Production	Merlin 50M	3 blade	B	32ft 2ins	29ft 11ins	Delivered as LF VB
Mk.VB	Production	Merlin 56	3 blade	B	36ft 10ins	29ft 11ins	Developed from Mk.I/II
Mk.VB	Production	Merlin 45	3 blade	B	36ft 10ins	29ft 11ins	Floatplane; 3 converted
Mk.VB	Production	DB605A	3 blade	B	36ft 10ins	29ft 11ins	EN830 captured
Mk.VB	Modified	Merlin 45	3 blade	none	36ft 10ins	29ft 11ins	ES127 conv 2 seater Italy
Mk.VB	Production	Merlin 46	3 blade	B	32ft 2ins	29ft 11ins	Delivered as LF VB
Mk.VC	Production	Merlin 45	3 blade	C	36ft 10ins	29ft 11ins	Developed from Mk.I/II
Mk.VC	Production	Merlin 45M	3 blade	C	36ft 10ins	29ft 11ins	Delivered as LF VC
Mk.VC	Production	Merlin 46	3 blade	C	36ft 10ins	29ft 11ins	Developed from Mk.I/II
Mk.VC	Production	Merlin 50	3 blade	C	36ft 10ins	29ft 11ins	Developed from Mk.I/II
Mk.VC	Production	Merlin 50A	3 blade	C	36ft 10ins	29ft 11ins	Developed from Mk.I/II
Mk.VC	Production	Merlin 50M	3 blade	C	36ft 10ins	29ft 11in	Delivered as LF VC
Mk.VC	Production	Merlin 56	3 blade	C	36ft 10ins	29ft 11ins	Developed from Mk.I/II
Mk.VC	Experimental	Merlin 46	3 blade	C	36ft 10ins	29ft 11ins	AB488 modified engine
Mk.VC	Production	Merlin 45M	3 blade	C	32ft 2ins	29ft 11ins	Delivered as LF VC
Mk.VC	Production	Merlin 50M	3 blade	C	32ft 2ins	29ft 11ins	Delivered as LF VC
Mk.VC	Production	Merlin 55M	3 blade	C	32ft 2ins	29ft 11ins	Delivered as LF VC
Mk.VC	Modification	Merlin 46	3 blade	2x MG	36ft 10ins	29ft 11ins	Delivered as HF VC
Mk.VI	Prototypes	Merlin 47	4 blade	A	40ft 2ins	29ft 11ins	X4942 as proto HF
Mk.VI	Production	Merlin 47	4 blade	C	40ft 2ins	29ft 11ins	100 built as HF
Mk.VI	Modification	Merlin 47	4 blade	deleted	40ft 2ins	29ft 11ins	Conv for Comms HF
Mk.VII	Prototype	Merlin 61	4 blade	C	40ft 2ins	31ft 3ins	AB450 converted
Mk.VII	Production	Merlin 61	4 blade	C	40ft 2ins	31ft 3ins	Mk.VI upgraded
Mk.VII	Modified	Merlin 61	4 blade	C	36ft 10ins	31ft 3ins	Normal wingtips
Mk.VII	Production	Merlin 64	4 blade	C	40ft 2ins	31ft 3ins	Late production HF
Mk.VII	Production	Merlin 71	4 blade	C	40ft 2ins	31ft 3ins	Final prod HF

Marque	Status	Powerplant	Propeller	Armament	Span	Length	Remarks
Mk.VII	Modification	Merlin 71S	4 blade	C	40ft 2 ins	31ft 3ins	LOX tests EN465
Mk.VII(PR)	Modification	Merlin 45	3 blade	A	36ft 10ins	29ft 11ins	Armed PR aircraft
Mk.VII(PR)	Modification	Merlin 46	3 blade	A	36ft 10ins	29ft 11ins	Was Type G
Mk.VIII	Prototype	Merlin 61	4 blade	C	40ft 2ins	31ft 4ins	HF wing JF299
Mk.VIII	Production	Merlin 61	4 blade	C	36ft 10ins	31ft 4ins	Mk.VII improved
Mk.VIII	Production	Merlin 63	4 blade	C	36ft 10ins	31ft 4ins	Main production
Mk.VIII	Production	Merlin 63A	4 blade	C	36ft 10ins	31ft 4ins	Late production
Mk.VIII	Modification	Merlin 61	4 blade	C	32ft 2ins	31ft 4ins	Far East clipped
Mk.VIII	Modification	Merlin 61A	4 blade	C	32ft 2ins	31ft 4ins	Clipped wings
Mk.VIII	Modification	Merlin 63	4 blade	C	32ft 2ins	31ft 4ins	Clipped wings
Mk.VIII	Modification	Merlin 63A	4 blade	C	32ft 2ins	31ft 4ins	Clipped wings
Mk.VIII	Production	Merlin 63	4 blade	C	40ft 2ins	31ft 4ins	Non standard trials
Mk.VIII	Production	Merlin 61	4 blade	C	36ft 10ins	31ft 4ins	Mk.23 wing JG204
Mk.VIII	Modification	Merlin 61	Various	C	32ft 2ins	31ft 4ins	JG424 Rotol Ltd
Mk.VIII	Production	Merlin 66	4 blade	C	36ft 10ins	31ft 4ins	LF production
Mk.VIII	Modification	Merlin 66	4 blade	C	32ft 2ins	31ft 4ins	LF production
Mk.VIII	Modification	Merlin 66	4 blade	C	32ft 2ins	31ft 4ins	Ground attack a/c
Mk.VIII	Production	Merlin 70	4 blade	C	40ft 2ins	31ft 4ins	HF production
Mk.VIII	Modification	Merlin 61	4 blade	C	36ft 10ins	31ft 4ins	Mod for comms
Mk.VIII(T)	Conversion	Merlin 66	4 blade	deleted	36ft 10ins	31ft 4ins	MT818/G-AIDN
Mk.VIII(T)	Conversion	Merlin 66	4 blade	optional	36ft 10ins	31ft 4ins	Service standard
Mk.VIII	Modification	Merlin 61	Contra	C	36ft 10ins	31ft 4ins	Trail a/c JK535
Mk.IX	Conversion	Merlin 61	4 blade	C	36ft 10ins	31ft 4ins	Mk.V improved
Mk.IX	Conversion	Merlin 63	4 blade	C	36ft 10ins	31ft 4ins	Mk.V improved
Mk.IX	Conversion	Merlin 63A	4 blade	C	36ft 10ins	31ft 4ins	Mk.V improved
Mk.IX	Conversion	Merlin 61	4 blade	C	36ft 10ins	31ft 4ins	Mk.V improved LF
Mk.IX	Conversion	Merlin 63	4 blade	C	36ft 10ins	31ft 4ins	Mk.V improved LF
Mk.IX	Conversion	Merlin 63A	4 blade	C	36ft 10ins	31ft 4ins	Mk.V improved LF
Mk.IX	Experimental	Merlin 63	contra	deleted	36ft 10ins	31ft 4ins	Rotol prop MH874
Mk.IX	Production	Merlin 61	4 blade	C	36ft 10ins	31ft 4ins	E-type wing
Mk.IX	Production	Merlin 63	4 blade	C	36ft 10ins	31ft 4ins	E-type wing
Mk.IX	Conversion	Merlin 61	4 blade	C	32ft 2ins	31ft 4ins	Clipped wings
Mk.IX	Conversion	Merlin 63	4 blade	C	32ft 2ins	31ft 4ins	Clipped wings
Mk.IX	Production	Merlin 70	4 blade	C	36ft 10ins	31ft 4ins	HF version
Mk.IX	Production	Merlin 70	4 blade	C	36ft 10ins	31ft 4ins	HF version E wings
Mk.IX	Conversion	Merlin 61	4 blade	C	36ft 10ins	31ft 4ins	Type A SAAF
Mk.IX	Conversion	Merlin 66	4 blade	C	36ft 10ins	31ft 4ins	Type B SAAF
Mk.IX	Conversion	Merlin 70	4 blade	C	36ft 10ins	31ft 4ins	Type C SAAF
Mk.IX	Conversion	Merlin 66	4 blade	deleted	36ft 10ins	31ft 4ins	Civil operated T/T
Mk.IX (PR)	Conversion	Merlin 70	4 blade	deleted	36ft 10ins	31ft 4ins	PRU conversion
Mk.IX(T)	Conversion	Merlin 66	4 blade	optional	36ft 10ins	31ft 4ins	Two-seat trainer
PR.X	Production	Merlin 77	4 blade	none	36ft 10ins	31ft 4ins	Pressurised Mk.XI
PR.X	Modification	Merlin 77	4 blade	none	40ft 2ins	31ft 4ins	As Mk.X, HF wings

Marque	Status	Powerplant	Propeller	Armament	Span	Length	Remarks
PR.XI	Production	Merlin 61	4 blade	none	36ft 10ins	31ft 4ins	Mk.IX modified
PR.XI	Production	Merlin 63	4 blade	none	36ft 10ins	31ft 4ins	Mk.IX modified
PR.XIII	Production	Merlin 63	3 blade	4xMG	36ft 10ins	29ft 11ins	Conv Mk.V for PR
Mk.XVI	Production	Merlin 266	4 blade	C	36ft 10ins	31ft 4ins	Standard wing-high back
Mk.XVI	Production	Merlin 266	4 blade	C	32ft 7ins	31ft 4ins	Clipped wing-high back
Mk.XVI	Production	Merlin 266	4 blade	C	36ft 10ins	31ft 4ins	Standard wing-high back
Mk.XVIe	Production	Merlin 266	4 blade	C	32ft 7ins	31ft 4ins	Clipped wing-low back
Mk.Ib	Conversion	Merlin 45	3 blade	D	36ft 10ins	29ft 11ins	Mk.V hooked
Mk.Ib	Conversion	Merlin 46	3 blade	D	36ft 10ins	29ft 11ins	Mk.V hooked
Mk.Ib	Conversion	Merlin 45	3 blade	D	32ft 2ins	29ft 11ins	Clipped wings
Mk.Ib	Conversion	Merlin 46	3 blade	D	32ft 2ins	29ft 11ins	Clipped wings
Mk.II	Conversion	Merlin 32	4 blade	deleted	36ft 10ins	29ft 11ins	Mk.I conv L1004
Mk.IIC	Production	Merlin 45	3 blade	C	36ft 10ins	29ft 11ins	Naval Mk.VC
Mk.IIC	Production	Merlin 46	3 blade	C	36ft 10ins	29ft 11ins	Naval Mk.VC
Mk.IIC	Production	Merlin 45	3 blade	C	32ft 2ins	29ft 11ins	Clipped wings
Mk.IIC	Production	Merlin 46	3 blade	C	32ft 2ins	29ft 11ins	Clipped wings
Mk.IIC	Experimental	Merlin 50	3 blade	deleted	36ft 10ins	29ft 11ins	LR694 Rotol tests
Mk.IIC	Production	Merlin 32	4 blade	C	36ft 10ins	29ft 11ins	Modified to LF
Mk.IIC	Production	Merlin 32	4 blade	C	32ft 2ins	29ft 11ins	LF clipped wings
Mk.IIC	Production	Merlin 55M	3 blade	C	36ft 10ins	29ft 11ins	Fixed wing F.III
Mk.IIC	Production	Merlin 55M	3 blade	C	32ft 2ins	29ft 11ins	Clipped wings
Mk.IIC	Production	Merlin 32	3 blade	C	36ft 10ins	29ft 11ins	FR conversion
Mk.IIC	Production	Merlin 32	3 blade	C	32ft 2ins	29ft 11ins	FR clipped wings
Mk.III	Conversion	Merlin 32	4 blade	C	36ft 10ins	29ft 11ins	MA970 as proto
Mk.III	Production	Merlin 55	3 blade	C	36ft 10ins	29ft 11ins	Mk.IIC folding wings
Mk.III	Production	Merlin 55	4 blade	C	36ft 10ins	29ft 11ins	As above
Mk.III	Conversion	Merlin 55	4 blade	C	36ft 10ins	29ft 11ins	Irish Air Corps
Mk.III	Conversion	Merlin 32	4 blade	C	36ft 10ins	29ft 11ins	Modified to LF
Mk.III	Conversion	Merlin 55M	4 blade	C	36ft 10ins	29ft 11ins	Bomb capable
Mk.III	Conversion	Merlin 32	4 blade	C	36ft 10ins	29ft 11ins	Modified to FR
Mk.III	Conversion	Merlin 55M	4 blade	C	36ft 10ins	29ft 11ins	Modified to FR

Armament Key:

A = 8x0.303 machine gun
B = 2x20mm cannon
C = 4x20mm cannon
D = 4x0.303 machine gun +2x20mm cannon

Engine Data

The Marvelous Merlin

Version	Max Power (hp)	Weight (lbs)	Installation	Remarks
C	1,045	1,370	K5054 initial fit	Development of P.V.12
F	1,045	1,410	K5054 engine change	Development of Merlin C
II	1,030	1,440	Spit Mk.I K9787-9960	Development of Merlin F
III	1,030	1,440	Spit Mk.I from K9961	As II with Universal shaft
IIIm	none given	1,400	Speed Spit K9834/N.17	III mod for high speed record
XII	1,150	1,450	Spit II only 1,104 built	III with Coffman starter
XX	1,240	1,450	Spit III experimental	Gear reduction to 42:1
32	1,645	1,430	Seafire IIc/III Spit XIII	Take off output increased
43	1,470	1,385	Spit V, PR.IV, VII Sea Ib/IIc	Improved design, rated at 16,000 ft
45M	1,585	1,385	Spit LF.V	45 rated for low level use
46	1,415	1,385	As per Merlin 43	Rated at 19,000 ft
47	1,415	1,400	Spit HF.VI	120 built High-altitude rated plus supercharger
50	1,470	1,390	Spit F.V, PR.IV	Similar to 45
50A	1,470	1,390	As before	As before
50M	1,585	1,385	Spit LF.V	Low altitude rated, 729 built in all
55	1,470	1,385	Spit PR.IV, F.V Sea F.III	Modified 50
55M	1,585	1,400	As before	Similar to 45M
56	1,565	1,400	As before	Similar to 55, 28 built
60	1,565	1,630	Spit III trial	For Wellington bomber
61	1,565	1,640	Spit F.VII, VIII, IX, PR.XI	Modified 60 reduction gear 42:1
63	1,650	1,645	Spit F.VIII, IX PR.XI	61 with modified supercharger
63A	1,710	1,645	As before	63 modified
64	1,710	1,665	Spit F.VII, PR.X	61 with cabin supercharger 182 built
66	1,580	1,645	Spit LF.VIII, LF.IX	61 modified for production purposes
70	1,475	1,640	Spit HF.VIII, IX PR.XI	66 with modified supercharger
71	1,475	1,650	Spit HF.VII	70 with modified supercharger 16 units
77	1,475	1,640	Spit PR.X	Modified cabin supercharger
266	1,580	1,645	Spit XVI	Packard built 66 gear ratio 0.479:1

SIGNIFICANT DATES

KEY DATES IN THE HISTORY OF THE SUPERMARINE SPITFIRE

1931
First Flight of Supermarine S.6Bs
Won 12th and final Schneider Trophy in September 1931.

20 February 1934
First flight of Type 224 (Spec F7/30).

3 March 1935
Contract 361140/34 issued for Type 300 to Spec F.37/34.

5 March 1935
Maiden flight of Type 300 K5054.
Later named Spitfire.

June 1937
Contract B817241/36 issued for Speed Spitfire, Type 323.

May 1938
First flight of production Spitfire Mk.I K9787 to Contract B527113/36 Type 300.

4 August 1938
Spitfire Mk.I to No. 19 Sqdn.

13 October 1939
Two Spitfire Mk.Is delivered to Heston for conversion to PR.IA.

January 1940
Six Spitfire Mk.Is delivered to Farnborough for conversion to PR.IB, Type 353.

16 January 1940
Spitfire PR.IB delivered to PRU.

13 February 1940
Spitfire Mk.II ordered to Contract B19713/39 (751 as Mk.IIA, 170 as Mk.IIB, Type 329).

March 1940
Spitfire Mk.III, Type 330/348, undergoes evaluation. Later converted to Griffon prototype.

22 March 1940
Contract B19713/39 for Spitfire Mk.V issued. Supermarine Type 331/349/352, over 6,500 built.

April 1940
Spitfire PR.IC delivered to PRU.

June 1940
Cannon-armed Spitfire Mk.IB delivered to No. 19 Sqdn.

August 1940
First Spitfire Mk.IIA delivered to No. 611 Sqdn.

October 1940
Spitfire PR.ID enters RAF service.

Late 1940
Spitfire IIC (ASR), Type 375, enters service with Nos. 276 and 277 Sqdns.

February 1941
First Spitfire Mk.V delivered to No. 92 Sqdn.

December 1941
First Spitfire Mk.VI rolled out.
Built to Contract B19713/39.
97 produced.

April 1942
No. 616 Sqdn receives Spitfire Mk.VI.

June 1942
First production Spitfire Mk.IX built to Contract B19713/39
Total built 5,117, Type 361/378

June 1942
First service Spitfire Mk.IXs delivered to No. 64 Sqdn.

June 1942
Spitfire Mk.VIII, Type 359, 360, 368, 376. Ordered to Contract Aircraft/1877/C.23(C). Total built 1,654.

August 1942
140 Spitfire Mk.VIIs ordered to Contract B19713/39.

November 1942
Spitfire PR.XI enters RAF service.
476 built to Contract B197813/39.

March 1943
Spitfire Mk.VII delivered to No. 124 Sqdn.

21 April 1943
Spitfire PR.XIII enters RAF service.

30 September 1943
Type 367/1 issued for Spitfire PR.XIII.

4 April 1944
First delivery of Type 387, Spitfire PR.X. Sixteen aircraft delivered to Nos. 541 and 542 Sqdns as part of Contract B19317/39

September 1944
Spitfire Mk.XVI deliveries begin.
1,055 aircraft built.